Surviving the Spreadsheet Tsunami

(Be Safe!)

RB Weber

Surviving the Spreadsheet Tsunami

The definitive guide to recognizing, addressing and preventing the flood of inaccurate, inefficient and ineffective spreadsheets that inundate almost every organization.

R B Weber

Well Worn Works Press

Houston, Texas

Except where otherwise noted, all problems and situations in this book are fictious.
Any resemblance to actual problems or situations is purely coincidental.

Well Worn Works Press is an imprint of

R Weber Associates LLC

ISBN: 978-1-7340931-1-7 (Paperback Edition)

First Printing

To Dan Bricklin and Bob Frankston who started it

And to all those who later improved it

TABLE OF CONTENTS

Table of Figures

INTRODUCTION: THE UNSEEN MONSTER

I developed my first electronic spreadsheet in 1982. I was working for the accounting firm of (what is today) Ernst & Young. We had just acquired an Apple II, which came with VisiCalc installed. I took to it like a hippopotamus takes to water. With my background in math & physics, I was captivated by its ability to use formulas and crunch numbers without using a pencil and paper (which, being a good scientist and inherently lazy, I always hated).

My initial opus was a simple model that calculated my accumulated savings based on some assumptions about interest rates and how much I would make and put away each month. According to the model, by the time I retired I would be one of the richest men in the world! Sadly (or fortunately??), this did not turn out to be the case. It quickly became apparent that there were a lot of complexities in my real-life spending and savings habits that my model had failed to incorporate. But it did provide an important lesson about the perils and pitfalls one can encounter when building computer models. Perhaps Philander Johnson said it best:

> *"Never before in the history of mankind has it been possible to produce so many erroneous results in so short a time".*

This is exactly the problem that many organizations face today, though all too often they don't realize it. Developing a spreadsheet is deceptively easy: Learn a few of the common functions, grab some data from the numerous reports floating around, turn the crank and *voila*, the mysteries of the universe are magically revealed! (Well, at least the mysteries of this month's operating results.) Others in your organization, while outwardly heaping praise on your apparent brilliance but secretly envious, decide that they too should create their own spreadsheets and partake in the glory. Soon, spreadsheets of various forms and questionable lineage are everywhere.

The monster has hatched, and it is hungry!

There is an old saying: "If the only tool you have is a hammer then everything looks like a nail." I prefer to look at it another way: "If you can only master one tool then a hammer is a pretty good choice." Hammers are very versatile. They are often used with nails, but they can also be used with a variety of other items. They can push in; they can also pull out. They can be used to smooth a surface or to create dents. They can straighten things that are bent. They can bend things that are straight. And much more besides.

Like any tool, it takes training and practice to learn how to use a hammer correctly and efficiently. It also requires care to maintain its usefulness. It takes experience to know when to use it (and, more important, when not). The same holds for electronic spreadsheet software. Learning the basic structure, concepts and functionality is one thing. Understanding what problems to tackle, and how to put these elements together to solve a particular problem is altogether different. This is where spreadsheet creation becomes a craft and ultimately an art. And it is something that only comes with experience, with trying many different approaches on many different problems, and with making mistakes along the way. This is not to say that one should look for ways to make mistakes. (Though heaven knows they are easy to create and often devilishly difficult to find.). By exploring new approaches one learns what does and doesn't work. And by "work" I mean creating a tool that can be used to effectively and efficiently address a business problem for an extended period of time.

This is especially true today. And this is the central message of this book. A spreadsheet is a tool for addressing a business problem. The measure of its success is the extent to which it assists in developing the solution to the problem. I will repeat this several times in the next 300-odd pages.

Spreadsheet software has advanced tremendously since it was first introduced by Dan Bricklin and Bob Frankston in 1979. The spreadsheet as an accounting construct has been around almost as long as accounting itself. Originally, they consisted of rows and columns of numbers that were manually maintained on large ledger sheets. About the only "function" was the row or column totals (think "SUM")[1]. As enlightening as these were, the process was both tedious and error prone. With the development of electronic

---

[1] If you remember the old movies based on Charles Dickens novel "A Christmas Carol" and the scenes where Bob Cratchit was scribbling away with a quill pen. He was probably updating a spreadsheet.

spreadsheet software on the personal computer it became feasible to manipulate much more data, and perform additional functions (mathematical, statistical, trigonometric, logical, etc.).

Users[2] quickly developed an insatiable appetite for these tools and were consistently pushing the software to the limits of its capabilities. Thanks to Moore's law[3], the power of personal computers has increased exponentially since they were first introduced. The manufacturers of spreadsheet software have happily taken advantage of this situation to add both functionality and capacity to their products. With each new advance, users quickly abandoned the previous version and adopted the new software. This has led to a succession of industry leaders, with Microsoft Excel currently dominating the market. Exact numbers are hard to come by, but a recent survey I found indicated that roughly 76% of businesses use Excel. I believe the actual number is a lot higher. For this reason I will be using Excel examples throughout this book, though the concepts being presented are applicable to other electronic spreadsheet products with equivalent functionality.

The capabilities of the modern spreadsheet software, supported by the power of the modern personal computer, can be, and is, used to address an incredible range of problems. These range from simple (such as a personal finance ledger) to complex (business analytics and decision

---

[2] I have always found it interesting that the two industries that traditionally refer to their customers as "users" are information technology and illegal drugs.

[3] Moore's Law: The number of transistors in a dense integrated circuit (i.e., its processing power) doubles every two years.

models). In support of this, a vast amount of information has been developed and made available. Lectures and training classes can be found for virtually any education level. Numerous books, articles, web sites[4], videos, blogs, etc. have been created. As of this writing, a quick search on Amazon showed 9,629 books about spreadsheets, with approximately 7,660 of them specifically about Microsoft Excel.

So why does the world need another book about spreadsheets? Unfortunately, almost all of the books currently available fall into one of two categories:

- *Basic spreadsheet concepts*: These discuss the basic principles, structure and functionality that electronic spreadsheet software provides. These product-specific offerings give details about product features, capabilities and limitations, the functions provided and their parameters, how it can interact with other products, data formats, and a wealth of other useful information.

- *Specific problem information*: These discuss how the spreadsheet software can be used to address various classes of problems (analytics, financial analysis, reporting, modeling, decision support, operational planning, etc.). These include discussions of the underlying theories, examples and detailed explanations of the formulas used. They frequently include downloads of "shell" spreadsheets that the user can adapt to their specific needs.

[4] If you're interested, one of the best that I've seen is the series of YouTube videos by Leila Gharani of Xelplus. She shows how to use the Excel functions in innovative and totally uncommon ways. Well worth a look.

These offerings provide an excellent foundation for individuals wishing to learn how to use spreadsheets, either generally or to apply in specific situations. What these don't provide are insights into the when, where, how and why a spreadsheet can, or should, be used in a particular situation. And, just as important, how to manage the environments where spreadsheets are used.

Returning to the hammer analogy, for many individuals the spreadsheet is the primary (or maybe only) tool in their toolbox. This is not necessarily a bad thing. But not every problem can be solved, or solved well, with a spreadsheet. Attempts to apply spreadsheet software to the wrong kinds of problems are often inefficient, incomplete, inaccurate, or incomprehensible. Which is a bad thing.[5]

To be truly effective, the user must know not only the capabilities of the tool (be it a spreadsheet or a hammer), but also be able to look at a new problem and understand how the problem can be structured so that it is amenable to being solved with that particular tool. These are two important points. First, when attacking any problem, one must think through the strategy for solving it. Sometimes this is so simple and intuitive that it is done almost subconsciously. Other times the solution is not so obvious and can require many hours of head-scratching (real or virtual). Second, devising the strategy must be done in the context of the tool(s) that will be used. And third, once the solution is developed it must be correctly managed and maintained. This is where experience and guidance

---

[5] Yes, I know people who use spreadsheet software as a word processor. Ughh!

becomes invaluable. With practice, this problem evaluation process becomes easier, and the problems being solved become more complex. Furthermore, these techniques and the thought process behind them can be used with other tools besides spreadsheets.

This book is not going to teach you Excel or any other type of spreadsheet software. It is assumed you already have a good working knowledge of its functionality (though I briefly review of some of the subtler basics to ensure we are all using the same language, understand some of the quirks and pitfalls that can force the developer to take a particular approach, and to explain why I recommend doing things in a particular way). What this book will show you is how to look at problems through the lens of a spreadsheet's capabilities, how to determine when and where a spreadsheet is the appropriate tool, how to correctly design and build complex spreadsheets to maximize their usefulness, and (especially important) how to manage the spreadsheet portfolio.

Spreadsheets have become a fixture in almost every organization. But all too often they create as many problems as they are built to solve. Worse, it may not even be apparent that these problems exist. Multiple spreadsheets can provide multiple conflicting answers to the same questions. They can become difficult (or impossible) to update to reflect changing business requirements. The recipe for the "secret sauce" embedded in them can become lost and forgotten over time.

The tsunami has arrived.

It doesn't have to be that way. If spreadsheets are designed, built, and managed correctly they can remain valuable tools for an extended period. If not, they won't. It's that simple.

This book will show you how.

1. THE ROOTS OF THE PROBLEM

The purpose of a spreadsheet is to solve a problem.

This is accomplished by extracting, analyzing and presenting information that cannot be easily ascertained simply by looking at the data. The problem to be solved might be a business-critical decision, or it might simply be idle curiosity.

To extract the data you must first understand the problem. If not, the data you extract will at best be meaningless and at worst misleading.

Spreadsheets exist to solve problems.

As discussed in the Introduction, the focus of this book is not to teach you Excel or any other spreadsheet software. It is to show you how to recognize and address the issues that are limiting its value as a business problem solving tool. This book is not going to show you a lot of clever Excel hacks (though it does contain examples to elucidate the point). It is going to show you why organizations frequently get mired in a deluge of less-than-useful spreadsheets, and how to rectify the situation.

There are many factors that can result in ineffective spreadsheet portfolios: The people building or maintaining the spreadsheets may not understand the subtleties of the tool they are using. The resulting spreadsheets can be poorly designed or built. They can include subtle errors that are difficult to identify and thus go unresolved. Ill-conceived attempts to address these errors can create other problems that get propagated into other spreadsheets. And, frequently overlooked, the environment in which the spreadsheets operate can present a wide range of issues and challenges if the necessary disciplines are not present. All of these can lead to situations where additional spreadsheets are created to "fix" the problems or limitations of previous spreadsheets, which only exacerbate the situation. Over time the situation can grow until it is completely out of control. The tsunami has arrived.

Unlike the tsunamis we hear about on the evening news, organizations facing the spreadsheet tsunami may not even realize it's happening. All they know is that it's taking more and more time and effort to get the information they need. The promised gains in productivity have become anything but. Just like those other tsunamis, the results can be disastrous.

So how do we deal with the tsunami? Ideally, prevent it from happening in the first place. If the tsunami has already arrived then the causes must be addressed and corrected. This takes a lot more effort, and in some instances the damage cannot be wholly rectified. At the least, future catastrophes can be averted.

At its root, the tsunami is usually the result of two basic issues:

- The individual spreadsheets may not be designed and constructed in an optimal, or even minimally effective, manner. Or, the logic behind the design is lost over time and subsequent iterations include escalating deviations. This can lead to a cascading series of problems.

- The environment in which the spreadsheets are used is not well managed and can therefore exacerbate these problems.

Either of these causes can result in hours of wasted effort and frustration. Unfortunately, they are often found together. Regardless, both need to be examined and addressed. This book discusses both these issues in detail.

Issue 1: Design and Construction

To address the first issue it is necessary to first step back and consider two fundamental questions that must be answered:

- Do we really understand the business problems we are trying to address?
- Is a spreadsheet the right tool to solve these problems?

Just as with the hammer analogy, sometimes a spreadsheet is not the best tool for the job. Understanding the problem is the first absolute requirement when building effective spreadsheets. What data is available? How does the data relate to the expected final solution? How does it need to be transformed to provide valuable information that will lead to a solution? How does this information need to be presented? These are all questions that must be addressed before one even begins to think about spreadsheets. Sometimes the answers to these questions are obvious (or, as is more often the case, the spreadsheet developer thinks they are); sometimes they aren't.

Once the problem is understood. The second question must be addressed. There are numerous solution alternatives available (both computerized and not). There is a real danger in assuming that a spreadsheet is the best (or only) approach available and proceeding blindly down that path.

As powerful as they are, modern spreadsheet software programs have limitations. While they can do a lot for you automatically, that is not always what you want it to do. By its nature, a spreadsheet imposes structures that can sometimes get in the way. A large part of developing an effective spreadsheet solution depends on understanding the structure and how to use (or circumvent) the constraints it imposes.

As part of our examination of different problems and how to address them I will also discuss the “right” and “wrong” ways of doing things. I put these terms in quotations because there are many considerations. Sometimes a “quick and dirty” solution is OK. For example, if you’re making a one-time-use spreadsheet to solve a specific problem you probably don’t need to worry about defined-range naming conventions. But a word of warning: If it really is a one-time-use spreadsheet, delete it after you have what you want. Too often one-time-use spreadsheets become institutionalized, and those shortcuts will come back and bite you as maintenance issues!

There are entire volumes available describing what are euphemistically called “best practices” I hate that term, because it implies that there is a single best way to do everything. There are a lot of good practices (and a lot of bad practices). I will discuss many of these throughout the book as we examine different problems. And I will explain why things should (or should not) be done in a particular way.

Issue 2: Environment Management

Perhaps the best way to understand the issue of environment management is to briefly review the history of computer programming. There are many analogies between what happened with the original procedural programming languages and new interactive non-procedural facilities such as spreadsheet software.

For most of the history of information technology programming activities were performed by, well, programmers. Programmers were viewed as an isolated group of weird people who engaged in esoteric practices but were generally considered (correctly, as it turned out) ignorant of the business needs of the users of their creations. During these early days there was a lot of frustration on the part of the users, but also on the part of the programmers. The users felt that, despite large expenditures of time and money, they weren't getting what they needed. The programmers were tired of being abused because the users couldn't articulate what they actually wanted. A complex set of rules evolved over time as a sort of uneasy truce between the two camps. These included detailed design specifications, programming standards, and project management methodologies[6].

---

[6] Important Sidenote: Fred Brooks led the project team that developed the original operating system for the IBM System 360 line of mainframe business computer systems (OS/360). The project eventually cost over a billion dollars (in the 1960s!). He later described this effort as "the greatest project management failure in history". His book about the experience, "The Mythical Man-month", is one of the greatest project management books ever written. If you haven't read it, you need to!

Everything changed in the 1980s and 1990s when spreadsheet software and other PC-based productivity tools became widely available. The users no longer needed the programmers. They could produce reports and other cool stuff themselves. The movement started small but grew rapidly as the spreadsheet software and the PCs it ran on became ever more powerful.

Soon spreadsheets were everywhere, and some user organizations began to question why programmers were even still needed. The technology leadership teams complained about the lack of discipline in the "shadow IT organizations" and warned that the unmanaged proliferation of PCs would cause problems. The users ignored them and went on happily building ever bigger and more complex spreadsheets. The first rumblings of the tsunami had arrived, though most organizations didn't yet understand the danger.

As spreadsheets became more widespread and took on an ever-larger role in supporting organization and management processes, issues began to emerge:

- Redundant versions of the same spreadsheet.
- Multiple spreadsheets providing inconsistent answers for ostensibly the same questions.
- Unable to reproduce results from previous runs.
- Unable to verify results against other data sources.
- Unable to explain how certain results were derived.
- Increasing frequency of errors.
- Increasing difficulty identifying and correcting errors.
- Increasing commitment of resources required to maintain the spreadsheets.

To name a few.

I need to pause here and make something very clear: Spreadsheets are programs. When someone develops a spreadsheet he or she is engaged in a type of programming. Not the same as using a procedural programming language (and I will discuss how profound these differences are in Chapter 3), but programming nonetheless. Over the years developers using procedural programming languages have learned, through painful trial and error, that there are specific practices (some general, some specific to the language being used) that must be followed to prevent problems. While the intuitive nature of the spreadsheet and other user-centric tools made producing useful products easier and quicker, it has not eliminated the need for structure and discipline. When building a spreadsheet the developer needs the discipline to understand and follow spreadsheet-specific practices. If not, the result can be catastrophic.

I spent the first part of my career as one of those "weird programmer types". During those days I learned a tremendous amount, not only about what makes a good program, but also how those programs need to be managed. Many of those processes are as applicable today as they were then. In later chapters I will discuss the ones that are relevant to the world of spreadsheets and why they are important.

The Path Forward

To be clear, the central character in this book is not the tool (i.e., the spreadsheet) but the business problem, the environment in which it is being addressed, and how to effectively use the spreadsheet as the tool to address the problem. In each of the following chapters I will examine various aspects of a problem, dissect them, and then go to the tool chest to find the functionality that can best be used to solve the problem.

Before we go on, I want to make clear what is not covered in this book, and why:

- **Connectors & Collaboration** – All of the examples in the book are done using the PC (single user PC or laptop) version of Excel. The data presented is assumed to be directly accessible to the spreadsheet (it may be a tab in the spreadsheet itself, a tab in another spreadsheet, or it may be another file altogether). Connectors tend to be very application or organization specific. As such, it is difficult to discuss them in generalities (other than that they are another way to get data into a spreadsheet, which I will discuss in a later chapter). They also are only peripherally involved in the analysis of a problem. *The problem and its data* is what's interesting, not the tool used to access the data. In the same vein, I am not going to discuss any of the numerous collaboration tools that are now available for Excel and the other office productivity products. Again, these are a means to create a solution to a problem and are not central to analyzing and solving the problem itself. Believe me: I firmly believe in collaboration and recommend (and engage in) it frequently. Many times it is the diversity of perspectives that ultimately reveals the solution. But as I said above, here I am focused on the business problem and the solution, not the path we followed to get there.

- **Extensions & Add-ins** – There are also numerous tools that have been developed that work with (or on) Excel. These provide a wide range of capabilities not directly provided by Excel. As discussed above, these usually provide assistance after the problem has been analyzed and a solution strategy defined.

- **Macros** – The great power of Excel is that it is a non-procedural tool. This means you don't have to define a discrete series of executable steps (i.e., an algorithm) to obtain the solution. In this respect it is inherently different from traditional (procedural) programming languages[7]. To be sure, the spreadsheet software itself contains a lot of programming. But this is all executed behind the scenes where the user doesn't have to deal with it.
  Excel does provide a facility for the user to insert procedural program code where necessary (or desirable). This program code is contained in macros. A macro can be invoked automatically by an event (such as 'spreadsheet open') or manually (via a combination of keystrokes such as '<ctrl>+r'). Macros are written in VBA (Visual Basic for Applications), a common object-oriented procedural programming language. Since macros are actual program code, they can do just about anything that is allowed based on the user's rights and permissions. (Some of the earliest computer viruses were actually implemented as macros.)
  There are two ways to create a macro in Excel. The developer can open the VBA editor and write one from

---

[7] As mentioned earlier, I will discuss the fundamental differences between Excel and procedural programming languages at length in chapter 3.

scratch, or he/she can use the Excel "Record Macro" feature to capture a series of keystrokes. This can be useful in situations where a spreadsheet requires a user to repeatedly perform the same series of operations. Regardless of how it is created, the result is still VBA code (which the developer can then edit). I am not going to discuss creating macros and VBA programming, but at various points I may comment that "this might be a good place for recorded keystrokes."

One additional caution: While developing or using a spreadsheet it is inevitable that the user will occasionally make mistakes. Excel thoughtfully provides an "Undo" feature which will reverse the last several actions (exactly how many is a parameter you can specify). The "Undo" function will not reverse the actions performed by a macro. If a macro screws something up you must figure it out and fix it manually. *Caveat emptor*!

- **Data Management** – As organizations have become increasingly "digital", the volume of data available to support management decision-making processes has grown exponentially. Unfortunately, this often results in situations where the data itself is incomplete, inconsistent or inaccurate. This can be a huge problem for organizations, but is outside the scope of this discussion. For our purposes we will assume that the needed data is available and of good quality. Our challenge here is to make the best use of it.

Now that we've agreed on what isn't covered in this book, let's look at what is.

Before we can develop an effective spreadsheet solution to a problem we have to decide if a spreadsheet is the right approach. First, we have to examine WHY we should even consider using a computer in the first place, and in particular, WHEN spreadsheet software should be the tool used to solve it. Sometimes this is obvious; sometimes it isn't. In addition to being effective, a problem-solving tool must also be EFFICIENT (though probably not in the way you're thinking). If not, if the selected tool isn't going to yield some significant efficiencies, why even consider using the tool? This is the focus of **Chapter 2**.

In **Chapter 3** I present some basic spreadsheet concepts that will be central to our discussion. This is not a spreadsheet 101 course. The reader should already have a good working familiarity with (at least) spreadsheet concepts and (ideally) Excel. The purpose here is to ensure you understand why programming using spreadsheet software is different from programming using a procedural language, and understand the implications. Additionally, the chapter will explicitly define the terminology and basic spreadsheet structures to ensure we have a common understanding.

The ease of creating a spreadsheet can lull the developer into a false sense of complacency. The result is frequently subtle problems that are difficult to identify or (as often happens) are swept under the proverbial rug to be dealt with at some undefined future time, probably by a player to be named later. **Chapter 4** is about common pitfalls frequently encountered when developing or modifying a spreadsheet. These problems are often obscure and will frequently lead to extensive hair loss.

The next four chapters deal with the conceptual elements common to every spreadsheet:

- **Chapter 5, Design** – Spreadsheets need to be consciously and deliberately designed just like any other program. A clean overall design structure is critical to developing an effective spreadsheet.

- **Chapter 6, Data** – Data is what you have to work with. Understanding the data aspect of a problem is essential, and is usually the first step in being able to solve a business problem. It is also where numerous incorrect assumptions are often made that can lead to significant issues later.

- **Chapter 7, Transformation** – In a nutshell, the power of a spreadsheet lies in its ability to convert raw data into useful information. Useful information can range from something simple, like a total, to complex, like an optimized production schedule. Whatever the desired end result is, getting there will involve some form of data transformation[8].

- **Chapter 8, Presentation** – Spreadsheets only become useful when the result is presented. Until then it is just zeros and ones floating around inside the computer. To be effective, the result must be organized, formatted and displayed in a manner that will maximize the engagement and comprehension of the target audience. The key phrase here is "target audience". What is perfectly appropriate for an accountant is probably totally inappropriate for the board room. Proper presentation is critical. And, oh by the way, the "target audience" may not even be a person. It is perfectly OK to use a spreadsheet to create data that is used as input to another program.

---

[8] I try not to use the phrase "data manipulation" as this makes the auditors nervous.

Chapter 9 discusses the often-overlooked considerations necessary to make the entire spreadsheet file ("workbook" in Excel nomenclature) an effective, manageable tool in the context of the environment in which it will be used.

Many organizations have policies and standards that are expected to be followed developing programs. Mature organizations have analogous policies and standards that apply when creating spreadsheets. This is essential in large organizations where the original creator of the spreadsheet may not be the same person who ends up using or maintaining it. These won't guarantee that the spreadsheets are "good" (however you want to define that), but at least they give the person who has to maintain the spreadsheet a fighting chance when researching a problem.

There are two very basic principles that must be followed to develop an effective spreadsheet portfolio. The first, and most obvious, is that the spreadsheet must be designed, developed and maintained in accordance to generally accepted good practices. This is the primary theme presented in the first nine chapters of this book.

The second principle is more subtle. Simply stated, the developer of a spreadsheet must possess a deep understanding of the business problem for which the spreadsheet is being developed to address. This is difficult to learn and much more difficult to accomplish in practice. If you pay close attention you will see that this principle is actually woven into the first part of this book.

In **Chapter 10** I take a different perspective and focus specifically on the problem and how a careful analysis of the problem can lead to its being addressed by a spreadsheet. In this chapter I present a variety of actual problems of various kinds I have encountered, their unique challenges, the steps I followed in analyzing them, and the resulting spreadsheet solution.

The last chapter of the book, **Chapter 11**, briefly discusses some of the other products now available that expand on and extend the capabilities provided by the basic spreadsheet software, and why electronic spreadsheets aren't going away any time soon.

Fasten your seatbelts, the ride is about to begin!

2. THE FIRST QUESTION

"If the only tool you have is a hammer, then everything looks like a nail."

There is a tendency today to think that every problem (business or otherwise) can be solved using a computer. And the people that develop computer software (be it anything from office productivity software to social media websites) are more than happy to nurture this belief. Reading an email is too time consuming, so the computer will read it for us. Looking at a map is too difficult, so the navigation program simply tells us where to go ("turn right at the next intersection"). And the list goes on. What isn't clear is exactly why we have come to accept this situation. Is it laziness, or has the world become so complicated that we don't believe we are capable of dealing with it? Or is it just nice to have something to blame when things go wrong?

Why (Use a Computer)?

Why do we need to use electronic spreadsheet software to solve business problems? Before we can answer that we need to ask an even more basic question: Why use computers at all?

The simple answer is because there are things that humans do well, and things that we don't do well. Likewise, there are things computers do well and other things that they don't. If it seems like I'm belaboring the obvious I am doing that for a reason: Understanding the fundamental nature of a problem is essential to picking the right tool to solve it.

For example, humans can solve pattern recognition problems far more easily than computers. This is why many online applications use CAPCHA or reCAPCHA to ensure the "user" isn't a robot. These tests typically present the user with a graphic or a collection of pictures, and then ask the user questions about it. Identifying which of a dozen random pictures contain an automobile is easy for you and me. For a computer to do that would require massive amounts of data and computation. On the other hand, something as trivial as adding a column of numbers is easy for a computer. You and I would probably make a mistake.

Some problems are best solved by combining the capabilities of both. For example, a typical cluster analysis problem might involve analyzing a large collection of data, associating values to some of the attributes, and then identifying the concentrations and outliers. A computer can certainly do the initial data crunching easily, but identifying the clusters is considerably more difficult (computationally intense). A good way to approach this problem is to let the computer crunch the data and then use it to produce a graph such as an X-Y Scatter Diagram. A simple glance at the resulting picture can immediately tell the user where the cluster(s) are, how concentrated, are there any outliers, etc. (It turns out I did exactly that for a major retailer that was having promotional merchandise stocking issues. This is one of the examples I discuss in Chapter 10) There are also an array of products now available that allow a user to work graphically and interactively with structured computer data such as Excel or database tables. I will touch on this in the Chapter 11.

While there are clearly tasks that humans can do better than computers, there isn't anything (at this point, at least) that computers can do that humans can't do (even if slowly and poorly). There is a story that Sir Isaac Newton hired a team of mathematicians to manually compute the areas contained in various shapes and curves as a cross check when he was developing calculus. Since most of us don't have the time or resources available to Sir Isaac Newton we have to rely on a different approach. That's where the computer comes in.

The computer provides us with a vehicle for doing certain tasks far more quickly and accurately than we can perform them manually. Modern electronic computers emerged almost 80 years ago. Their development was largely driven by the needs of the military during the second world war, and following that the cold war. Since then, their capabilities have grown significantly (in both speed and storage capacity) but they still function in essentially the same way[9].

Consequently, the computer's great strengths are the same now as they were then:

- They can process large volumes of data efficiently
- They don't make mistakes (or they are at least honest about it when they do).[10]

If the characteristics of the problem to be addressed involve substantial amounts of processing or substantial quantities of data then the problem is most likely a good candidate for a computerized solution. If it involves pattern recognition or evaluating culturally significant human responses maybe not.

Step 1 complete: We have determined that the problem under consideration should be solved on a computer. The question then becomes: which computerized solution?

---

[9] For those of you who are interested, they use what is called the "fetch/execute cycle". The program instructions are stored as data in the computer memory. A special piece of hardware referred to as the program address register tells the hardware to read the contents of a specific memory location ("fetch"), interpret it, and perform the indicated action ("execute"). Then the program address register is updated, and the cycle repeats.

[10] This is not to say that computer programs don't have errors in them, and that the results produced are always correct. A significant part of the circuitry in a modern computer is dedicated to preventing, identifying and correcting internal data errors. What people do with the data is not the circuitry's problem.

When (To use a Spreadsheet)?

There are many computerized problem-solving tools available. Which one should be used is a critical decision. Each tool is best suited for a particular type of problem. Choosing the wrong one can result in much wasted time and effort, or complete failure.

Figure 1 (below) provides a 10,000-foot view of certain problem data characteristics to consider when deciding which tool is appropriate for a particular problem.

FIGURE 1. TOOL SELECTION

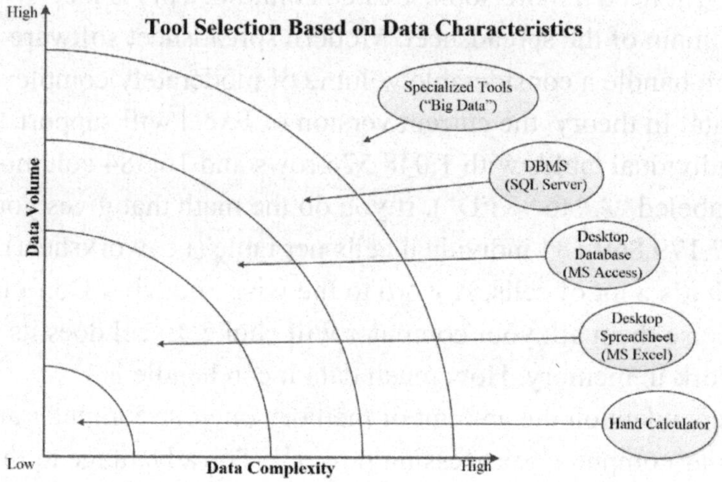

This chart shows the tool that would be appropriate for solving a problem as a function of the volume and complexity of its data. Data volume refers to the number of records and elements in each. Data complexity refers to the number of record types, the number of relationships, the degree that the data is structured, and other considerations. Note that these are rough approximations. The drawing is certainly not to scale, and there is considerable overlap in the capabilities of the various tools. There are other factors that also might be considered.

In the bottom left corner are the problems with a small volume of simple data. You don't need a spreadsheet or a database to determine whether the regular or the jumbo box of cereal on the shelf in the grocery store has a lower unit cost per ounce. The calculator app on your smart phone (or in your head) is more than adequate.

If the problem is somewhat larger or more complex you might need a more sophisticated solution. This is the domain of the spreadsheet. Modern spreadsheet software can handle a considerable volume of moderately complex data. In theory, the current version of Excel will support individual tables with 1,048,576 rows and 16,384 columns (labeled "A" to "XFD"). If you do the math that gives you 17,179,869,184 individual cells per table (i.e. worksheet). That's a lot of cells. A word to the wise, though – Don't try to use them all; your computer will choke. Excel does its work in memory. How much data it can handle is dependent on the amount of memory on your computer and your computer's processing power[11]. The advantage to this approach is speed. Excel does not have to move data into and out of the disk storage. It can therefore perform rather complex calculations involving a large number of cells quite quickly.

---

[11] Just for a giggle I tried to load a worksheet with "0" in every possible cell on a tab. My computer choked.

For problems that are somewhat larger than what Excel can conveniently handle there are desktop database applications such as Microsoft Access. Access can handle substantially larger problems because the Access tables are stored on disk. When Access needs to perform an operation it reads the required records from disk, updates the data, and writes them back. The size of an Access database is therefore constrained by disk capacity rather than memory. The tradeoff is speed, since disk input/out operations (I/O) is significantly slower than memory accesses.

The primary constraint on a desktop database systems is processing power. With the ability to access external storage (either local hard drives or on a network) the amount storage available is theoretically unlimited.[12] But all of the database activity must go through the desktop computer's processor(s). Though far greater now than in the early days of desktop computing, capacity is still finite. These databases are also typically limited to a single user at a time.

Problems that are larger and more complex, or have additional requirements (such as concurrent multi-user access, robust backup, security, activity logging, rollback capabilities, etc.), require server-based database management systems. These systems frequently employ dozens (or hundreds) of servers, terabytes of storage, and can handle millions of records, thousands of connected database tables, and thousands of concurrent users. The

---

[12] The current (2016) version of MS Access has a limit of 2 gigabytes per database file. However, this limit can be circumvented by linking tables in multiple database files.

three systems most frequently found in large organizations today are Microsoft SQL Server, Oracle DBMS, and IBM DB2 (for IBM mainframes).

Everything we have looked at up to now involves what is usually referred to as "structured data". This is data that can have multiple occurrences of multiple record types, but each record type has some pre-defined collection of data elements. This is the kind of data most often encountered in financial and other typical business applications.

Beyond the domain of the industrial-strength database management systems lies the domain of the special purpose systems. For example, one such domain is known as "big data". These systems are designed to handle huge volumes of frequently unstructured data streams. For example, systems of this type do the front-end processing for each of the experiments at the CERN Large Hadron Collider physics laboratory in Switzerland. These systems analyze and evaluate roughly a terabyte of data streaming from several hundred thousand individual electronic detectors every second during a continuous three-year experiment run. That's a lot of data.

Another special purpose domain are those programs that work with "soft" data, such as audio and video files. You can store soft objects such as pictures in a spreadsheet file, but the spreadsheet can't actually do anything with it other than display it.

We have already decided that the problem to be addressed has characteristics that make it amenable to a computerized solution. Now we need to decide which one. If these characteristics include a moderate number of records of a moderate number of structured record types, and will only be used by one user at a time, then the electronic spreadsheet is a good candidate.

I deliberately chose the term “moderate” because it is somewhat ambiguous. How big (in terms of records and types) your system can handle is entirely dependent on the capabilities of your system. And to be honest, the only way you’re going to know for sure is to try it. That said, modern systems can do a lot. I have a laptop with an Intel i7 quad-core processor and 8 gig of memory and it can handle spreadsheets with 100,000 records and about a dozen record types without any problems.

Efficiency

The reality of the world we live in today is that we are drowning in data, both as individuals and as members of our various organizations. Our sanity, if not our survival, depends on our ability to process this flood efficiently. This situation is frequently the fault of our computers, but our computers also give us the means to efficiently manage at least some of these challenges.

The operative word here is efficiency. Simply defined, efficiency means making the best use of the most expensive resource. Once upon a time, computers were very expensive (in addition to being big and slow), and programmers were cheap. (I know – I was one of them.) In those days efficiency meant teaching the computer to do the maximum amount of work using the minimum storage and processing cycles. Programs were written in computer assembly code to optimize every instruction. The fact that the average output of an assembly language programmer was on the order of three to five lines of debugged code per day was immaterial. Today the situation is reversed: computer processing power is abundant and cheap, and the developers (we've graduated from being mere programmers) are expensive.

So efficiency now means taking advantage of the cheap and abundant processing power to minimize the work that we expensive humans have to do. The entire nature of information technology has evolved in response to the changes in the *system economics*.

This is where electronic spreadsheet software can be a lifesaver and a game-changer. Spreadsheets provide the ability to accept large quantities of data (how to accomplish that is discussed later), perform various calculations and other transformations (again, later), to distill the data into valuable information that can be presented in a user seductive (user friendly is so last year!) format. And (this is key) and do all this *efficiently*. Efficiency is the second most important consideration, the first obviously being accuracy.

What exactly are we concerned with when we talk about efficiency? As I explained above, it isn't the computer anymore, it's the people. Specifically, it's the people who:

- Originally develop the spreadsheet
- Use the spreadsheet to solve specific business problems
- Maintain the spreadsheet as requirements change

In some situations these roles are all performed by the same individual. In others they can be performed by different people entirely. This is frequently the case with spreadsheets that support activities in large organizations. Some efficiency considerations are common to all three roles; some are specific to each.

CONSIDERATIONS WHEN DEVELOPING A SPREADSHEET

When we open Excel to begin developing a spreadsheet we are presented with a blank pallet. Excel gives us a framework – a set of labeled cells arranged in rows and columns. The developer takes this pallet and uses it to create tables, reports, and other structures. He or she will add formulas to transform the data into useful information, and formats to present the information in a pleasing fashion.

Developing a spreadsheet <u>is</u> programming. But it is quite different from traditional programming using a *procedural* programming language like (originally) FORTRAN or COBOL or (more recently) object oriented languages like Visual Basic, C++, or a host of others. In a procedural programming language the developer starts by determining **what** the desired result is, and then tells the computer, step by step, exactly **how** to achieve it. Excel and other spreadsheet software products are *non-procedural.* The developer still must determine the **what**, but the spreadsheet software decides **how**. A huge portion of the development process has been eliminated!

The other great benefit is that spreadsheets can be developed on a personal computer by the person most familiar with the problem instead of needing a large, expensive mainframe and a team of technologists. The invention of electronic spreadsheet software allowed, for the first time, people without formal training in a programming language to create their own "programs." Spreadsheet software is generally credited with moving the personal computer from the hobbyists' garages into the business office. As discussed above, the primary efficiency consideration today is to optimize the use of the expensive humans by leveraging the availability of cheap and abundant computer processing power. Electronic spreadsheet software exemplifies this shift in perspective.

Please note: Efficiency in developing a spreadsheet is not a license for sloppiness. There are three important reasons for this:

- First, the developer must never lose sight of the fact that spreadsheets, especially in large organizations, may be used and/or maintained by someone else. Sloppiness on the front end can create huge inefficiencies (read: hours of problem solving) on the back end.

- Second, during the development process one will almost certainly encounter errors – formulas that don't work, results that don't look right, boundary conditions that weren't anticipated, etc. Poor development practices can make the process of "debugging" the spreadsheet significantly more difficult.

- Third, even though computer processing power is cheap and abundant, it is finite. Too much sloppiness and garbage will bog down performance.

Conceptually, the developer of a spreadsheet should follow many of the same "good programming practices" that developers using traditional programming languages follow. (As mentioned previously, large organizations often have standards or policies that spreadsheet developers are expected to follow.) Some of the more common recommendations include:

- **General Structure** – Like any program, a spreadsheet should have a clean, well defined structure. The details will vary from application to application, but in general, the logic should follow a "linear" flow. This is discussed in detail in Chapters 5 and 9.

- **Table Layout** – As a general rule, each worksheet within a spreadsheet should contain a single table/record type (if data) or report. I will talk about report design and layout extensively in a later chapter. There are several considerations that can make the development and maintenance of data tables more efficient:
    - Tables are almost always laid out with each row representing an individual record, and each column a unique data element within the record. This is why Excel gives you over a million rows but "only" 16,000 or so columns. If this orientation is not the case, there should be a good reason for it.
    - Sometimes it is most expedient to include formulas in a data table. For example, this might be done to create logical "flags" to indicate specific data conditions on each record, to add a lookup value to the data, or to create data elements (through concatenation, truncation, etc.) that are not native to the inputted data. When this is done keep the original data separated from the derived data; do not intersperse them. I usually keep the inputted data on the left and the derived data to the right, separated by a blank column.
    - Each data element needs to be identified. Put a descriptive header at the top of each column of the table to define the meaning of the data elements in that column. If necessary, use multiple rows to describe data groupings, identify which columns contain original data and which contain derived data, etc.
    - One of the most powerful tools that Excel provides is the "Data Filter". Every data table should have filtering enabled to facilitate finding specific

records. I usually insert a blank row below the headings and use it for the filter row.
- If the data table is going to be used as input to a PivotTable, Excel requires that the top row of the table contain a unique identifier in every column of the table. This includes blank columns. For reasons I will discuss in the next chapter, these identifiers should be descriptive, but as short as possible. I frequently hide this row to avoid confusion.
- Be sure to freeze the header row(s) so they don't roll off the screen when scrolling through the data.

➢ **Ranges** – Many spreadsheet functions require a range (a rectangular collection of contiguous cells) as input. In a large spreadsheet the same range may be used multiple times (for example, as input to multiple reports). As the data grows or changes it is imperative that the ranges that reference that data also grow and change. It is also imperative that all the functions or reports that use that data access the same range of cells. Inconsistencies in range definition (or references to individual cells) is one of the most common causes of problems in spreadsheets. To avoid these problems (and the time wasted trying to find them) the developer should do the following:
- All commonly used ranges should be given meaningful names. Naming a range in Excel is extremely easy – just go to the address bar and type the name. Any function or activity that requires that range of cells should use the assigned name, not the cell reference.
- Since data tables often grow over time, it is often desirable to leave extra empty rows at the bottom of the table to allow for growth. Include these empty rows in the definition of the range.

- Identify where the range ends in the table. A common technique is to put a comment ("End of Table <whatever>") in the row directly below the bottom left hand corner of the range, and fill the cells to the right with a character (such as "x") to highlight the bottom of the range. This will also serve as a visual check if for some reason the data should overflow the defined table range. (This is very helpful when refreshing a PivotTable, as Excel will warn you that the cells already contain data. Then you know there is a problem.)
- If it becomes necessary to expand a table (either rows or columns), do the insert within the bounds of the range. This will automatically expand the range. If you do the insert at the edge of the range (last row or last column) it won't, and the additional data won't be picked up when the range name is referenced.

➢ **Freezing Cells** – In Excel, a reference to a cell can be "frozen" by placing a $ sign immediately in front of the row or column reference (or both) of the cell. What this does is prevent the cell reference from changing as the formula using that cell reference is copied. If the cell reference isn't frozen, the row refence advances as the formula is copied down, or the column reference advances as the formula is copied across. Sometimes this is desirable; sometimes it isn't. As the developer it is your responsibility to determine which. If done wrong, you can spend a lot of time trying to figure out why your formulas give goofy results. Some other considerations:

- When you name a range of cells, Excel automatically freezes the cell references that define

  - the range. Therefore, if you copy a formula using a named range the range definition doesn't change.
  - Anytime a cell has a specific meaning, any reference to that cell should be frozen.
  - As a general rule, it is preferable to freeze the cell definitions in a formula unless there is a specific reason not to (i.e. the formula is going to be copied and each copy must reference different cells)

- **Formulas** – The true power of spreadsheets lies in its ability to easily evaluate extremely complex formulas referencing many individual data points. Creating and debugging those formulas is often not so easy. Consider the formula in the following screenshot:

```
=IF(ISTEXT($M120),0,IF(Z$2<$M120,0,IF(ISNUMBER($Q120),ROUND(MAX(0,MIN(IF(ISNUMBER($P120),$P120,99999),IF(ISNUMBER($O120),$O120*Z$3,99999),IF(ISNUMBER($N120),$N120*
VLOOKUP($Q120,Avail_Hrs,Z$1+3),99999),IF(ISNUMBER($S120),$S120-SUM($V120:Y120),99999),BG120,IF(ISNUMBER($L120),VLOOKUP($L120,DepTbl,Z$1+1),99999))),0),0)))
```

Yeah, I can't figure it out either, and I wrote it. It is the formula to determine how many hours a specific resource can work on a specific task in a specific period. It is part of a project planning / resource scheduling spreadsheet. I will talk about this spreadsheet in detail in a later chapter. For now let's just consider this one formula. Even if it was larger it would still be difficult to understand. It is quite complex, with many embedded conditions. It would be much easier to understand if each element of the formula was split out. Something like this:

FIGURE 2. BREAKING UP A COMPLEX FORMULA

```
=
IF(ISTEXT($M120),0,
IF(Z$2<$M120,0,
IF(ISNUMBER($Q120),
ROUND(MAX(0,
MIN(
IF(SNUMBER($P120),$P120,99999),
IF(ISNUMBER($O120),$O120*Z$3,99999),
IF(ISNUMBER($N120),$N120*VLOOKUP($Q120,Avail_Hrs,Z$1+3),99999),
IF(ISNUMBER($S120),$S120-SUM($V120:Y120),99999),
BG120,
IF(ISNUMBER($L120),VLOOKUP($L120,DepTbl,Z$1+1),99999)
)),0),0)))
```

The formula is still complex, but now at least the individual elements are identifiable and can be analyzed individually to determine if each is performing as expected.
To do this, insert a <new line> character (keystroke <alt>+<enter>) into the formula where each logical break occurs. This does not affect the formula, only the way it is displayed in the formula bar. Note that you may have to expand the formula bar to see the entire formula.

All the considerations discussed above concern the spreadsheet itself, and guidelines that should be followed to make the spreadsheet easier to understand, debug, use, and maintain. But what about the spreadsheet development process itself? One measure of development efficiency (in any programming language) is accomplishing the most with the least number of keystrokes. Excel provides a number of ways to do that:

- **Quick Access Toolbar** – The Quick Access Toolbar (QAT) lies at the top left of the Excel screen. It contains buttons that perform, with a single click, a variety of actions. Each of these actions can also be performed by selecting the appropriate ribbon tab, clicking on the the correct (though often not obvious) icon, choosing the desired item from the dropdown menu, etc. But why go through all this if you don't have to?
  The default configuration of the QAT as Excel comes out of the box has five "buttons" – Save, Save As, Undo, Redo, Print. You can (and should) customize the Toolbar to present buttons for actions that you perform frequently. This is a screenshot of how I have my QAT configured:

I've arranged the buttons in (what I consider) logical groupings. From left to right they are:

- File control – File Save, File SaveAs
- Display control – Freeze/Unfreeze Panes
- Editing – Undo, Redo
- Print control – Set Print Area, Page Setup, Print Preview, Print
- Formula debugging – Evaluate Formula, Trace Dependents, Trace Precedents, Remove All Arrows
- PivotTable Control – Toggle Field List, Refresh All
- Filter Control – Set Filter, Clear Filter Selection

These are actions I use frequently. Having them as buttons on the QAT has saved countless hours of hunting through the ribbons.

- **Status Bar** – The Status Bar (SB) is at the bottom of the screen. It shows various file and screen status information. It can also be configured to show various arithmetic functions performed on the currently selected range of cells (Sum, Count, Average, Min, Max, etc.) This can also be helpful when testing or debugging the spreadsheet.

- **Other Keystroke Savers** – Excel follows the normal Windows conventions, so common functions such as cut, copy and paste can be performed using the standard Windows shortcuts (<ctrl>+x, c and v respectively). In addition, shortcuts exist for many of the Excel functions. Some of these are specific key sequences; others are implemented on the pre-defined function keys. I have included a partial list of these in the Appendix for your reference. Experience will teach you which you use most frequently and are therefore worth remembering.

One final word to the wise: As discussed previously, Excel does everything in memory. Therefore, it is important that you take steps to protect your work so you don't lose it. This is especially true if you are using a PC that doesn't have an internal battery or isn't connected to a UPS.

Excel does provide an "Autosave" capability. I prefer not to use it. I like to save my spreadsheets at logical points in the development process rather than at a specific time interval. But keep in mind that once the file is saved there is no going back. It is not unusual to find you've made changes and later determine that they aren't working as intended.

To avoid these problems it is always a good idea to create **versions** of your spreadsheets. This way, if you need to fall back, you have previous versions to go back to.

What could be more inefficient than having to reenter hours of work?

CONSIDERATIONS WHEN USING A SPREADSHEET

So what do we mean by "using" a spreadsheet? For our purposes I will define "using" to mean entering or updating the data tables, updating the results provided by the formulas that use the data, and using the updated data or formula results to produce a report. Let's look at each of these individually:

- **Entering or updating the data** – Depending on the specific spreadsheet and environment, this could consist of keying values into specific cells, importing the data from a file or another spreadsheet, or executing a data connection.

- **Updating the results** – Usually Excel updates (recalculates) the values of formulas each time a cell is modified. Excel is smart enough to only recalculate the values in cells that use the modified cell (i.e. are dependent on it). There are some situations where this doesn't happen:
  - Recalculation mode is set to "Manual". This is sometimes done if the spreadsheet has a large number of formulas using a large amount of data and recalculating on every change would be too time consuming. In this situation the user must force Excel to recalculate when the updated results are needed. Function key <F9> will do this.
  - PivotTables must be updated manually by invoking the "Refresh" function.
  - Sometimes you need to do managed iterations to get an approximation via a "trial and error" type of "Goal Seek" process. Excel provides facilities for

doing this, but they have to be invoked manually.

- **Produce an updated report** – Once the data is entered and all formulas or structures recalculated, the "reports" in the spreadsheet will reflect these changes and can be used however they were intended.

Efficiency considerations when using a spreadsheet primarily consist of exercising caution and using the spreadsheet as it was developed for its intended purpose. Key things to pay attention to include:

- Ensure the data is entered correctly and only into those areas where intended. The Excel Protection feature (discussed in Chapter 9) can be used to enforce this.
- Check to ensure that no tables have overflowed.
- Update all formulas (recalculate, if necessary), PivotTables (refresh), etc.
- Save the updated spreadsheet as a new version with identification as to why, when and by whom the update occurred.
- Save prior versions in some kind of "archive" folder so they can be referenced later if needed, and so they aren't confused with the most recent version.

The penalty if these aren't done will be either having to do it over or providing inaccurate information. Neither of these are desirable.

CONSIDERATIONS WHEN MAINTAINING A SPREADSHEET

Spreadsheets have a nasty habit of evolving over time. Requirements change. Could be a new report. Could be additional data. Could be new rules. And as they do, the spreadsheets must also change to incorporate these new requirements. As mentioned earlier, often these changes are implemented by someone other than the original developer of the spreadsheet. Here is where adhering to good development practices (such as identified above) can reduce the number of expletives hurled at the developer or rude comments made about his or her ancestry.

All of the suggestions provided above also apply when maintaining the spreadsheet. In addition, there is one cardinal rule that applies specifically to maintainers: If something is no longer needed, **GET RID OF IT!**

The worst situation that a person maintaining a spreadsheet can encounter is a lot of old, no longer used, data or formulas. I have spent numerous hours debugging spreadsheets where previous developers have hidden sheets or left formulas that no longer contribute to the creation of the reports, only to conclude that they should be ignored.

If the requirements have changed, or for whatever reason sections of a spreadsheet are deemed obsolete, either create an entirely new spreadsheet or comment that the new version does not include the stuff that is being removed and why. Keep the prior versions so that they can be referenced or retrieved if necessary. But definitely keep the latest version as clean as possible.

Your fellow developers will thank you for it.

3. ESSENTIAL FUNDAMENTALS

Before we begin exploring the problems and practices that can result in an unmanageable spreadsheet portfolio I feel it is important to review some of the spreadsheet fundamentals. Some of the material presented in this chapter may seem redundant since it has been stipulated that the reader should already be familiar with spreadsheet software.

I feel that this review is important to ensure that we have a common understanding of the elements of a spreadsheet, and the terminology used to identify and describe them. There are also some subtleties in the way certain things work that are often overlooked. These can be the source of problems that are difficult to find and correct.

More than that, I want the reader to really understand why spreadsheet software is fundamentally different from other programming languages. It is this fundamental difference that makes spreadsheets so powerful. Understanding this difference is central to understanding what spreadsheets can, and can't, do.

The New Paradigm

When Dan Bricklin and Bob Frankston invented the first spreadsheet program they created, perhaps without even realizing it, a radically new programming paradigm. Spreadsheet programs operate in a way that is profoundly different from the way traditional procedural language programs work:

- In a procedural language program the data is defined explicitly as part of the program and can be different in each program. In a spreadsheet the data is predefined and fixed. Every worksheet using the same version of the software has exactly the same number of data elements (i.e., cells) in each tab, whether or not they are all used.

- In a procedural language program the code and the data elements are separate and distinct. The commands to create the data and the code to act on the data are almost always in different parts of the program. In a spreadsheet the data and the code (i.e., the formulas) are together in the same place (the cell).

- In a procedural language program the order in which the instructions are executed is generally determined by their position in the program: Line 56 in the program is generally executed after line 55 and before line 57. In a spreadsheet the order of execution is determined by the hierarchy of the cell references. If cell C6 has a formula that uses the value of cell J3, cell J3 is evaluated before C6.

- In a procedural language program the code "pushes" values to the individual data elements For example: In a

COBOL[13] program there might be this statement – MOVE “STUFF” to DATA-STORAGE (where “DATA-STORAGE” is a data location whose name and characteristics were explicitly defined in the DATA DIVISION). In a VBA program the same statement would be – DATA-STORAGE := “STUFF”. In a spreadsheet, values are “pulled” from one cell to the next based on the cell references.
This last difference is **extremely important**, as it constrains what can and can’t be performed in a spreadsheet. This will be discussed further in Chapter 9.

[13] COBOL??? Yes, I’m showing my age.

Terminology

The first requirement for a common understanding (be it a discussion about spreadsheets or life in general) is that we have a common terminology. I've listed some terms below to clarify how the various elements of an Excel spreadsheet fit together. I will be using this terminology for the remainder of the book:

- **Environment** – All of the examples I use are based on Excel running on a Windows operating system. Within Windows, the location of a specific file is identified with a letter (often "C:"[14] for the internal disc drive of the PC), followed by a series of hierarchically defined file directories.

- **Program** – Excel is a program that has the filename Excel.exe on a computer that is running the Windows operating system. Spreadsheets are not programs in the traditional sense in that they don't contain machine (or compile to) executable machine instructions. They are data (and in the newer versions of Excel they are data stored as XML[15]). They are interpreted by the Excel program to create on the screen what you see as a spreadsheet.
  That said, once you are in Excel the act of creating or modifying a spreadsheet <u>is</u> (as was emphasized earlier) programming and needs to follow the appropriate rules and disciplines. I know this may feel confusing but deal

---

[14] The early versions of the IBM PC had two floppy drives which were labeled "A:" and "B:". The internal (hard) disk drive, which came later, was subsequently identified as the "C:" drive. This nomenclature has persisted long after the demise of the floppy.

[15] And they can be viewed or edited with an XML editor. I don't recommend this.

with it.

- **Spreadsheet** – This is probably the most confusing, and misused, term. For our purposes, I will refer to the spreadsheet as a single file, which may contain a collection of tabs. This will be clarified in the following.

- **Workbook** – To avoid this confusion, Excel refers an Excel file (i.e. a file with an extension of .XLS, .XLSX, .XLSM, etc.) as a workbook. For our purposes, I will use the term "workbook" when referring to the physical file and the term "spreadsheet" when referring to the logical or conceptual entity.

- **Worksheet** – Excel (and most other current spreadsheet software) allows the creation of multiple tabs within a workbook. It refers to each of these as a worksheet. It is important to be cognizant of this, for example, when analyzing the "scope" covered by a named range. – a named range can be local (applies to only one worksheet), or global (applies to the entire workbook). As above, I will use the term "tab" to refer to the physical construct in the workbook and worksheet when discussing the logical construct.

- **Cell** – Each worksheet (tab) in an Excel spreadsheet (workbook) consists of a collection of cells organized into rows and columns. The rows are identified with a number (1 through 1,048,576) and the columns are identified with a letter ("A" through "XFD") in the current versions (2016 or later).
  Every cell in a Worksheet has contents which yield a "value" of some kind (more on this later). A cell can be empty (<NULL>), which as it turns out <u>is</u> a value and can be used in various functions. The value of a cell can be a simple number (e.g. 6) or a simple text (e.g.

"Hello") or the result of a formula (e.g. =SUM(A17:G43)). Formulas are identified by the first character being an equal ("=") sign. This tells the Excel program that it needs to interpret the contents of the cell to obtain the value.

THIS IS IMPORTANT!

You should always think of a cell as having two personalities: The one is the **value** returned by the cell when it is viewed or referenced, the other is the **contents** of the cell that resulted in that value. For example, if you do a "Find" (<ctrl>+F), you have the option to search for a cell whose contents contains the character string you are searching for, or to search for a cell whose returned value is the character string that you are looking for. Which option you select can yield very different results.[16]

---

[16] THIS IS IMPORTANT, TOO: Auto Filter always searches for and selects cells based on their value. More on this later.

Specific Elements

Numbers – The value of a non-empty cell can be one of two general types – It can be text or a number. As a general rule Excel doesn't try to do any interpretation with text. On the other hand, numbers can mean many different things. Excel does its best to accommodate this by allowing us to format the number when it is displayed in a manner consistent with its intended usage. Internally, Excel stores numbers using whatever capabilities the hardware and operating system provide. When a number is displayed Excel interprets the number based on the direction the developer has provided (or its best guess if no direction has been provided):

- **Number Formats** – Excel provides several different default number formats based on the intended interpretation of the value of a specific cell. For example, if the developer indicates that a collection of cells is to be interpreted as currency, Excel will automatically format them with two decimal places (for US dollars), use commas (",") for the thousands, and add the appropriate currency symbol ("$") in front. Other frequently used formats include Percentages, Fractions, Scientific (i.e. powers of 10), special sequences for things like social security numbers, or General (which means Excel will try to display it in the best way it can given the value it finds). You can even define exactly how a number is displayed by writing your own format specification using standard format codes.[17]

[17] For convenience I have provided a list of the standard format codes in the Appendix.

- **Date and Time** – Dates and times are stored as numbers just like any other. The value of a date is calculated as the number of days since January 1, 1900. Therefor the date October 21. 2019 is stored internally as 43,759 (the number of days since 01/01/00). Times are stored as a decimal fraction of 24 hours. 6:00am on October 21, 2019 would be stored as 43,759.25. This makes date and time arithmetic extremely easy, and Excel provides a variety of functions for doing just that.

Formulas and Functions – The great power of Excel is its ability to support and evaluate extremely complex formulas which may contain many functions. When discussing these formulas and functions, it is important to be consistent in how we identify the cells that are used or impacted:

- **Target Cell** – The target cell is the cell that contains the formula that will be evaluated and the resulting value displayed when the cell is referenced or selected.

- **Source Cell (s)** – The source cell(s) are the cells that provide the values that will be used as the parameters to the function(s) contained in the target cell.

Excel provides a wide range of built-in functions[18]. All Excel functions require some type of parameter (even if it's a null "()" parameter) and return a value. The returned value can either be text or numeric. The number and type of parameters required, and the type of value returned depend on the specific function. The important thing to note here is

[18] And you can create your own functions. These can be written in VBA using the macro editor. As I mentioned earlier, I'm not going there.

that, since functions return a value, and functions require values as parameters, a function can be a parameter to another function. In other words, there is nothing wrong with "nesting" functions within functions within functions within functions.... The value supplied as a parameter to a function can be coded directly in the formula, the contents of a specified source cell (or range of cells), or the result of another function.

Consider the following simple example of a formula in a certain target cell using the VLOOKUP function:

```
=VLOOKUP(C6,Desc_Tbl,3,FALSE)
```

The VLOOKUP function has four parameters:

1. The value being searched for. In this example, the value is the <u>evaluated contents</u> of source cell C6 (which may itself contain another formula).
2. The range of source cells that contain the table of data being searched. In this example, the range has been given the name "Desc_Tbl"
3. The offset (column) in the data table that has the desired data element. In this example, that element is found in the third column from the start of the table.
4. An optional parameter specifying whether the search requires an exact or approximate match. In this example, the supplied value of FALSE indicates an exact match is required.

Cell Reference – Many formulas refer to a cell as a parameter to a function. The underlying question (to Excel, at least) is where that cell reference refers to. There is a hierarchy of possible answers:

- If the reference is a named range, the name explicitly identifies where the cells are located. The exact definition of the range can be found by invoking the Name Manager, which is located on the Formulas tab of the Ribbon (or by clicking <ctrl>+F3). The name tells Excel everything it needs. No further explanation necessary. Be careful, though. Name definitions can be local (applies only to the current tab) or global (applies to the entire workbook). If local, the same name can appear in different tabs with different definitions.

- If the cell reference is to a cell in the current active worksheet then the cell address is sufficient. As in the above example:

  C6

- If the cell reference is to cell in another worksheet (tab) in the same file (workbook), the reference will be identified as <tab name>!<cell reference>. If the tab name has spaces then <tab name> is enclosed in single quotes. For example:

  'Customer Info'!C6

- If the cell reference is to a cell in another file (workbook), the reference will be '[<complete file name including path>]<tab name>!<cell reference>. For example:

  '[C:\Today's Data\Input File.XLSX]'Customer Info'!C6)

The danger of referring to cells in another spreadsheet is that the reference is very dependent on the environment and file directory structure. What works on one PC may not work on another. For this reason, the ideal is for all references to be contained within a single workbook (file) or on a common (shared) file storage device.

Before or After – Sequence is a basic concept in both our lives and in our computers. Sequence has two possible meanings:

- As a noun, sequence refers to the order in which a collection of items is arranged. The concept of sequence is extremely helpful when we search for a specific item in a list. We learn the letters of the alphabet in a specific order so when we are looking for something that begins with the letter "P" we know it is going to follow the items that begin with the letter "O".

- As a verb, sequence refers to the act of arranging a list of items in a specific order. This is the underlying concept behind the act of sorting.

In Excel, and in computer software in general, there is an intrinsic sequence associated with items based on the "code" (i.e. collection of bits) used to represent them. Laptops and Personal Computers generally use what is known as ASCII (American Standard Code for Information Interchange). In ASCII, Each letter, number, symbol, special character, etc. is assigned a specific ASCII code comprised of seven binary bits. For reference, the complete list of ASCII codes is shown in the Appendix.

In many computer programs, the codes assigned to a set of items determines the sequence of the items (based on whether the binary value associated with one item is greater or less than that of another). Excel does not strictly follow this convention. First, the sort is based on the evaluated **value** of the cell, not its **contents**. This can make a difference if the cell contains a formula. Second, Excel uses an internal coding structure that results in cells sorting in the following order:

- Numbers based on their numerical value from most negative to most positive (Note that this includes dates and times.)
- <space>
- Special Characters in order based on their ASCII code
- Letters in alphabetical order, with lower case preceding upper case (i.e. a,A,b,B,c,C,…y,Y,z,Z)
- <NULL> (i.e. an empty cell)

Sequence is extremely important, as it can directly impact the way certain functions operate. In the VLOOKUP example shown above, if the user instead specified TRUE for the final parameter Excel would try to do an approximate match. For an approximate match to work correctly the table being searched <u>must</u> be in ascending key value order.

It is also important to note that table entries that are empty (have a <NULL> value in the key field) will sort behind (below) the entries that have non-null entries. In other words, when a table is sorted Excel squeezes all the empty records to the end.

CURIOUSER AND CURIOUSER!

One of the parameters you can specify in the sort command is what "direction" to perform the sort. In other words, numeric fields can be sorted either lowest to highest or highest to lowest. Text fields can be sorted A-Z or Z-A. If two cells containing text values are identical except the first character of one is lower case and the first character of the other is upper case, the lower case row will always sort first, regardless of which direction was specified. Empty cells behave in a similar manner – they will always sort to the end regardless of direction.

The following screenshot shows a table with various values sorted on the first column lowest to highest:

FIGURE 3. EXCEL SORT SEQUENCE

Excel Sort Sequences		
Value	**Description**	**Cell ASCII Code (Octal)**
-1	Negative One	N/A
0	Number Zero	N/A
1	Positive One	N/A
38	Formula (=counta)	N/A
4/16/2020	Date	N/A
	<blank>	040
!	Exclamation	041
""	Double Quotes	042
#	Hash	043
$	Dollar	044
%	Percent	045
,	Comma	054
.	Period	056
@	At	100
a	Lower Case "a"	141
A	Upper Case "A"	101
z	Lower Case "z"	172
Z	Upper Case "Z"	132
	<NULL>	N/A
TRUE	Logical True	N/A
FALSE	Logical False	N/A

True or False? – In addition to numbers and text, an Excel formula can include logical values. These are values that are treated as either TRUE or FALSE. Excel also has functions that work on TRUE and FALSE values similar to how arithmetic functions work on numeric values. The logical functions supported by Excel are: AND, OR, XOR, and NOT. Note: These are implemented as functions and have to be coded like any other function. Consider the following examples. In each example, cells A1 and A2 contain a logical value (either TRUE or FALSE):

=A1+A2 ← Legal. The result is numeric, not logical.

=A1 AND A2 ← Not legal. Excel will give an error.

=AND(A1,A2) ← Legal. The result is a logical value.

How Excel evaluates a particular formula can be made dependent on whether an expression evaluates to either true or false. The Excel function "IF" is used for exactly this purpose, as in the following formula:

=IF(C6>C5,<Do this if TRUE>,<Do that if FALSE>)

Any numeric value can also be interpreted as a logical condition. If a numerical expression evaluates to zero then it is treated as FALSE when tested; if it evaluates to anything other than zero (positive or negative) it is treated as TRUE.

The following two formulas are equivalent (if both cells have numeric values):

=IF(C6-C5,<Do this if TRUE>,<Do that if FALSE>)

=IF(C6<>[19]C5,<Do this if TRUE>,<Do that if FALSE>)

Logical values can also be designated explicitly. The text strings "**TRUE**" and "**FALSE**" have special meaning and are reserved for this purpose. A cell can be assigned a value of TRUE or FALSE. When this is done the value stored in the cell will be one or zero, respectively. The cell can then be used in other expressions. The table below shows what you get if you add two cells containing logical values together:

FIGURE 4. SUMS OF LOGICAL VALUES

	A	B	C
1	**Sums**	**TRUE**	**FALSE**
2	**TRUE**	2	1
3	**FALSE**	1	0

THIS IS WHERE IT GETS WEIRD!

When an empty cell (i.e. evaluates to <NULL>) is used by itself it tests as FALSE. For example, if cell C6 is empty then in the following formula,

=IF(C6,<Do this if TRUE>,<Do that if FALSE>)

the FALSE action will be performed. **But**, if an empty cell is used with the AND logical function it is treated as TRUE. Using the same conditions as in the above example,

=IF(AND(C6,1),<Do this if TRUE>,<Do that if FALSE>)

the TRUE action will be performed. **But**, if an empty cell is

[19] The logical operator "<>" is the way Excel specifies "Not Equal". It is the opposite of the logical operator "=".

used with the XOR logical function it is treated as FALSE. Using the same conditions as in the above example,

=IF(XOR(C6,1),<Do this if TRUE>,<Do that if FALSE>)

the TRUE action will again be performed. In other words, it appears that an empty cell will do whatever it needs to do keep the logical functions happy. Go figure.

There are other ways in which the <NULL> value are treated in ways that may not be inherently obvious. These are discussed in the following sections.

Case (In-)Sensitivity – As was seen in the discussion on sorting (above), Excel does not treat upper and lower case letters in the manner one would expect. This idiosyncrasy (I can't think of a better word to describe it) permeates Excel in several ways:

- If the only difference between two strings of characters is that one has upper case letters and the other has lower case letters then they will be treated as being the same. This is true whether the strings are in cells or hard-coded into a formula. In the following formula:

  =IF("A"="a",<Do this if TRUE>,<Do that if FALSE>)

  the TRUE action will be performed. The same will occur if two cells are tested, or any other combination.

- When used as a search argument in a lookup function, Excel cannot distinguish between entries with upper and lower case letters. The screenshot below shows the result of several VLOOKUP functions using a portion of the

"Excel Sort Sequences" table that was shown previously:

FIGURE 5. VLOOKUP USING UPPER & LOWER CASE ARGUMENTS

	A	B	C	D	E	F
2		Excel Sort Sequences				VLOOKUP Using
3		Value	Description	ASCII Code (Octal)		Column B to Search
5		0	Zero	060		Zero
18		a	Lower Case "a"	141		Lower Case "a"
19		A	Upper Case "A"	101		Lower Case "a"
20		z	Lower Case "z"	172		Lower Case "z"
21		Z	Upper Case "Z"	132		Lower Case "z"
22			<NULL>	N/A		Zero

The VLOOKUP in column F uses the contents of column B as the search argument, and returns the value in column C. As you can see in rows 18 to 21, it doesn't matter if the search argument is upper or lower case, the result is always the lower case table entry because it is the first entry that matches the search criteria. If the entries were reversed, the VLOOKUP would have found the upper case table entry. Using a subset of the same table, but with the entries in the opposite sequence, the results are as follows:

	A	B	C	D	E	F
2		Excel Sort Sequences				VLOOKUP Using
3		Value	Description	ASCII Code (Octal)		Column B to Search
4			<NULL>	N/A		Zero
5		Z	Upper Case "Z"	132		Upper Case "Z"
6		z	Lower Case "z"	172		Upper Case "Z"
7		A	Upper Case "A"	101		Upper Case "A"
8		a	Lower Case "a"	141		Upper Case "A"
9		0	Zero	060		Zero

To Null or Not To Null – As discussed above, an empty cell (i.e. a cell with no **contents**) has a **value** of <NULL>. This is a special situation in a spreadsheet. There is actually no way to assign a cell a <NULL> value[20].

- The number zero is not treated the same as <NULL>.
- The text string "" is not treated the same as <NULL>.

A cell can be tested to determine if it is <NULL> with the ISBLANK function. ISBLANK is a logical function that returns a value of TRUE if the cell being tested is empty; otherwise it returns a value of FALSE.

One situation where <NULL> values are often found is when data is loaded into a predefined table space. If the table space is larger (i.e. more rows) than the data loaded into the table, the extra rows will have no data and be <NULL>.

Obviously, another way to empty a cell (or range of cells) is to select the cell(s) and press the 'Delete' key. The contents of the cell(s) are removed, and the value of the cell(s) are now <NULL>. When using a spreadsheet it is often good practice to empty all of the cells in a table space before loading new data. This prevents old data from corrupting the results if the new data being loaded has fewer rows than the previous data.[21]

---

[20] You can of course write VBA code to clear the contents of a cell and do it in a macro, but that would be cheating.

[21] This might be a good place for recorded keystrokes. Create a macro to delete the table data space, put a button at the top of the worksheet, and give it a name like "Clear Table".

As long as we're on the subject, notice in the VLOOKUP example above what happens when an empty cell is used as the search argument. It produces the same result as using zero. One more example of <NULL> having an identity crisis.

Defaults – Like most software, Excel has default values for most of its settings. Many functions and operations have optional parameters, and Excel needs to have a value to use if you don't specify one. Since there are literally thousands of settings in Excel, there is a good chance you are using many of them. Always check to determine what Excel is assuming about your functions and formulas to ensure it is doing what you expect. Often it is a good practice to explicitly set a value, even if the default is consistent with what you want. Sometimes default values change from version to version, and what used to work might not in the future.

Backward Compatibility – Once upon a time, Lotus 1-2-3 was the dominant spreadsheet software application on the Personal Computer. Lotus was bought by IBM in 1995, and since then it has largely faded into oblivion. In Lotus, keyboard functions were invoked with a "/" followed by a character string. In recognition of this, if "/" is the first character entered it has same effect as pressing <alt>. Excel tries to interpret anything following it as a command.

4. COMMON PITFALLS

The first absolute requirement of every spreadsheet is accuracy. Whatever the spreadsheet was designed to do, it must do it correctly. If not, why even bother? Problems occur because it isn't always obvious that the spreadsheet isn't accurate. Or, it may be apparent that something is amiss but the cause of the issue is obscure.

Overall, spreadsheet software does a good job of being "intuitive". You can look at the data in a table and see exactly what's there. You can click on a cell and get a pretty good idea of what it does by studying its contents in the formula bar. This is one reason why it has become so widely used by non-IT personnel.

But as with previous intuitive programming languages (remember "self-documenting" COBOL?) this is not always the case. As was shown in Chapter 3, sometimes the facilities provided by Excel don't work exactly as expected, or they have limitations that make them less than "user friendly." Or, everything in the spreadsheet may be working exactly as directed, but what you coded isn't exactly what you wanted. Or, maybe the data contains conditions you didn't expect and didn't allow for.

When building a complex spreadsheet a lot of time can be spent trying to figure out why a specific function or formula isn't working the way you expected or think it should. It is important to understand the not-so-obvious things that can cause Excel to behave oddly. In the discussion that follows I present some general situations and frustrations that are frequently encountered. Throughout the book I'll identify other specific "Gotchas!" as they are encountered.

What You See; What You Get

The image below is a screenshot of the upper left corner of a spreadsheet. Look at cells A3 to A7:

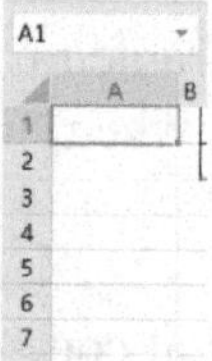

To all appearances, the cells are empty. However, if we expand the view we can see what is actually in each cell:

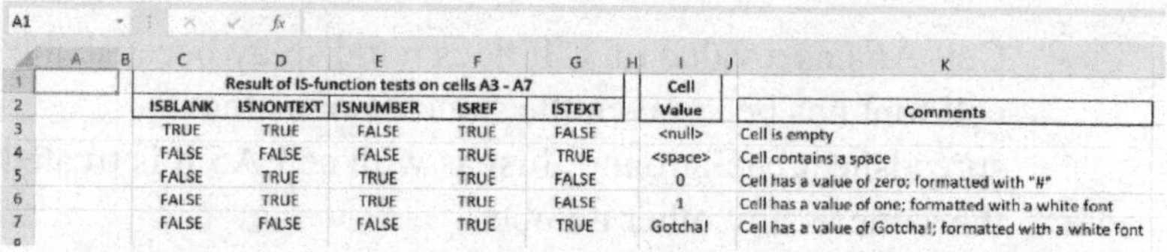

ISBLANK	ISNONTEXT	ISNUMBER	ISREF	ISTEXT	Cell Value	Comments
TRUE	TRUE	FALSE	TRUE	FALSE	<null>	Cell is empty
FALSE	FALSE	FALSE	TRUE	TRUE	<space>	Cell contains a space
FALSE	TRUE	TRUE	TRUE	FALSE	0	Cell has a value of zero; formatted with "#"
FALSE	TRUE	TRUE	TRUE	FALSE	1	Cell has a value of one; formatted with a white font
FALSE	FALSE	FALSE	TRUE	TRUE	Gotcha!	Cell has a value of Gotcha!; formatted with a white font

Result of IS-function tests on cells A3 - A7

Each of those cells has a different value. It is evident from the results of the various "IS-functions", Excel treats each very differently:

- Cell A3 is really empty (i.e. evaluates to <NULL>). This is shown by the ISBLANK function returning a value of TRUE. If you use this cell in an arithmetic formula it will be treated as if it has a value of zero.

- Cell A4 has a space. A space is treated as text (the ISTEXT function returns a value of TRUE). If you try to use this cell in an arithmetic formula it will return an error (#VALUE!).
  The cell could actually have several spaces in it. One

way to determine how many is to select (click on) the cell, then click on the right end of the formula bar and see where the curser positions. If it doesn't go to the extreme left then there is data in the cell.
Please note – this is not just a trivial exercise. Trailing spaces in a cell are not visually apparent but will cause functions such as VLOOKUP or HLOOKUP to not find the desired entry. Because of this it is often desirable to use the TRIM function to remove excess blanks when loading data into a table.

- Cell A5 has a value of zero. Because of the format code applied to the cell ("#") it displays as a blank. It is still treated as a number (the ISNUMBER function returns a value of TRUE). Even though it doesn't show on the display, it is treated the same as any other number.

- Cell A6 has a value of 1. It does not display because the cell font has been given the same color as the spreadsheet background. Just as with cell A5 it is treated the same as any other number.
  This trick can be useful when, for example, it is necessary to include a cell in a report (to drive a conditional format or as a field in a PivotTable) but you don't want the data to display. This is discussed in more detail in the chapter on presentation.

- Cell A7 has a value of Gotcha! It does not display for the same reason as cell A6. As with cell A4, it is treated as text.

As you can see from the above discussion, what a particular cell contains is not always obvious from its appearance. Often the best way to determine what a cell actually contains is to click on it, then click on the formula bar (as described for cell A4, above). Many times a formula not performing as expected or returning an unexpected error can be traced to a cell containing an unexpected value that is not apparent from simply looking at that portion of the spreadsheet on the screen.

Special Situations

Numbers vs. Text: A situation frequently encountered that can cause problems is when numbers are store as text. Sometimes a cell can contain a value that looks like a number but is intended to be treated as text. This can often happen when data is being pulled into a spreadsheet from an external source.

Consider the following example. This is part of the invoice table for a small manufacturing company. The data was imported to a spreadsheet from the company's accounting system:

	A	B	C	D	E	F
1	Customer Invoice Data					
2	Customer	Date	Invoice Nbr	Part Nbr	Amount	Description
4	1846-wjn	1-Apr	1896	863	126.49	Small Widget
5	1264-tfw	1-Apr	2114	864	417.83	Big Widget
6	2914	1-Apr	2123	217	42.86	Thigamajig
7	1846-rce	2-Apr	2261	149	33.18	Stuff
8	4141	2-Apr	2389	666	89.15	Junk

The table contains the Customer ID, invoice date, invoice number, part number, amount and part description. There is another table in the spreadsheet that has customer information (Customer ID, name, address, phone, etc.). The customer information table is used in reports and is accessed using the Excel VLOOKUP function.

As you can see, some of the Customer IDs have a three letter suffix (maybe the buyer's initials) and some don't. All of the Customer IDs are text values, even the ones that don't have the suffix. The Customer IDs in the customer information table are also text. If any of the Customer IDs that don't have a suffix are "touched" (i.e. clicked on), Excel will automatically convert them to the corresponding numeric value. If this happens, any lookups for that customer will fail and return a #N/A error.

Excel does provide a warning that a cell contains a numeric value stored as text. This is indicated for cells A6 and A8 in the above example by the small green triangle in the upper left hand corner. Unfortunately, Excel doesn't warn you if a cell that is supposed to have a text value doesn't. The best thing to do in this situation is include a column of cells to perform the ISTEXT test and use it to produce a warning. Alternatively, depending on the application, there are ways to force a cell or an entry in a table to be treated as text.

Hidden Treasures: There are many ways to hide rows, columns, or individual cells in a spreadsheet. There are also many reasons why you might want to. Some of these were discussed in the preceding section. The important consideration is to understand what happens when these things are hidden. Failure to understand this can lead to many hours of head pounding and hair pulling.

- **Zeroing Row Height or Column Width** – Any row height or column width can be set to zero. Doing this will prevent them from showing on the display but does not "hide" them (make them go away). All the cells are still there, and any actions that cover a range of cells that includes those rows or columns will affect those cells as well. This includes actions like copying formulas. The important exception is that the "Find" command will not

locate any cells in those rows or columns.
Sometimes a row height or column width can be set to zero inadvertently. If a column width has been set to zero and this isn't noticed while you are, for example, setting up a table you can end up with an extra blank column. When data is copied into the table the fields would be shifted and not align with the correct column headers. Similarly, if a row height has been set to zero a record could be copied in that won't be visible or searchable.
Note that any action to adjust the row height or column width on a range that covers these rows or columns will act on them as well, effectively "unhiding" them. Sometimes this is the easiest way to find and correct the problem.

- **Hiding Rows or Columns** – Excel allows you to explicitly hide rows or columns using the "Format" command. When this is done the results are similar to those as when the row height or column width is set to zero. If you manually select a cell (key the address into the address bar) in a hidden row or column and then invoke the <format><row/column><height/width> command you will see that is exactly what Excel did (i.e. set it to 0).

- **Filtering** – The data filter is a powerful Excel facility that allows you to selectively hide or unhide rows in a table based on the value(s) in a specific column. There are some important differences between filtering and the other hiding methods described above:
    - Filtering is cumulative, in the sense that the available filtering criteria for a column is limited based on the filters that have already been applied for other columns.

- Actions such as "Copy", "Paste", or clearing cells (selecting a range and pressing the Delete key) automatically exclude cells that are hidden by a filter.
- If you hide rows using one of the methods discussed above, the filter list will only show the values in the unhidden rows. **But** – if you make no changes to the filter criteria and click OK all of the hidden rows are automatically unhidden. If you select to hide additional rows those rows will also be hidden. If you click on the filter list again it will now show all of the hidden rows, but that they are now hidden by the filter, not by the <format><hide><row> or <format><row><height> commands.
  (Is your head hurting yet?)

➢ **Hiding Tabs (Worksheets)** – In my opinion, there is no good reason for hiding tabs. I have seen spreadsheets where the developer has hidden tabs that contain reference data to prevent them from being changed. That doesn't work – hiding is not the same a protecting. Use the "Protect" feature to protect things. I have seen spreadsheets where certain "secret" information (salaries, billing rates, etc.) that is used in certain formulas was put in a hidden tab. That doesn't work – if it can be selected it can be seen. If it can't be selected it can't be used, so why is it there at all? I have seen spreadsheets where maintainers have hidden tabs because they are obsolete. As I explained before, that is a criminal act and should be made punishable by the death penalty. If it is obsolete get rid of it (but keep previous versions in a separate workbook file with an appropriate explanation).

Ranges: The PivotTable is one of Excel's most powerful features. It can be used to create reports and a whole lot of other cool stuff (even data tables that can be used as the source for other PivotTables!). And it all begins with a range.

The first (logical) step in creating a PivotTable is to select a range of data. The first row of the range must have a unique identifier in each column, followed by rows of data. As every good developer knows, these ranges should be named. When the "Insert PivotTable" command is invoked, the selected range of data is automatically populated into the "Table/Range" field. **But** – Excel does not populate the field with the name of the range, it populates the absolute reference (rows/columns) of the range. What this means is that as the data range changes, the PivotTable doesn't change. Additional data won't be included; obsolete data won't be excluded. To avoid this situation, when creating a PivotTable it is important to override the "Table/Range" field with the range name.

More PivotTable Idiosyncrasies: In addition to creating reports, the PivotTable can be used as a powerful tool for selecting and organizing data. In other words, a PivotTable can take data from one table and use it to create another table which can then be used in subsequent formulas.

When creating a formula, the developer would typically select (click on) the desired cell at the appropriate spot in a table of data. When this is done Excel populates the formula with the relative reference of the selected cell. The formula can then be copied down, and the resulting formulas are applied to each of the rows in the table.

But – When a PivotTable result cell (i.e. one that is specified in the "Σ Values") is selected, Excel populates the formula with a "GETPIVOTDATA" function that specifies the criteria (i.e. row and column values) for that specific cell[22]. Think of it as an absolute cell reference for an entry in a PivotTable. When a formula with the GETPIVOTDATA function is copied the reference doesn't change. The resulting formulas will all select the same cell from the PivotTable. Sometimes this is what is desired, but usually it isn't. If a relative reference to a cell in a PivotTable is needed it has to be specified explicitly.

Note – if a cell is selected that is in the row or column specification portion of the PivotTable it populates with the correct relative cell reference.

Finding Faults: As discussed previously, each cell in a spreadsheet has two personalities: the contents of the cell (text, number or formula), and the resulting evaluated value of the cell. The "Find" command defaults to "formulas", which means it will search for a cell whose contents contain the specified character string. If a user wants to find a cell containing a specific value, the "Look in" option must be changed to "Values".

For example, a table might have a column that contains formulas with the VLOOKUP function. At the bottom of the table is a SUM of that column for all the rows in the

---

[22] Full disclosure: This is an option that can be turned off if desired (but seldom is). Go to the "Working with Formulas" section in File > Options > Formulas.

table. If the SUM has a value of "#N/A", that probably indicates at least one of the VLOOKUP functions failed (i.e. there was no entry in the lookup table that matched the specified search criteria). Since none of the formulas contain the string "#N/A" a Find command won't select any of the cells in error unless the option is changed to "Look in Values".

Insert or Delete: When inserting of deleting rows or columns some things adjust but others don't. Things that do adjust include:

- In general formulas adjust, even if they contain absolute cell references.

- Ranges (named or explicitly specified) adjust, but only if the insert or delete is done <u>between</u> the first and last row or first and last column.

- If columns are inserted (or deleted) in a table that is used as the source for a PivotTable, the PivotTable will pick up (or drop) the resulting field specifications, but only after the table is refreshed.

- Conditional formats do, but beware! If a conditional format references multiple rows or columns, Excel doesn't want to assume that you want the new row or column to have the same formatting. So it obligingly gives you the option. When inserting a row or column it splits the format range, lists the two parts separately, and creates a new conditional format entry for the inserted row or column. The result is you can quickly get incomprehensibly complex conditional format definitions. You need to periodically go in and clean them up.

Often a major source of errors in spreadsheets is due to the fact that Excel doesn't adjust lookup offsets. The VLOOKUP and HLOOKUP functions require an offset be specified to select the column or row (respectively) that contains the desired data. If columns or rows are inserted into or deleted from the table used in these functions the offsets in the formulas that reference that table do not automatically change. This will result in the wrong data being selected if it isn't fixed.[23]

Sorting Formulas: If you sort cells that contain formulas, the formula will change to maintain the same references relative to the new position of the cell, unless the cell references are absolute ($). In other words, sorting effectively performs the same way as copy/paste. When the cell is evaluated, it will have a new value. If you sort it again, the cells could end up in a different order based on the new value of the formula.

Calculations: Excel defaults to calculating (reevaluating) a cell, and all cells dependent on that cell, each time the cell is changed or accessed. Normally this good, because the data shown on the screen accurately reflects what is currently in the spreadsheet. This can be a problem for

[23] The 2020 version of Excel introduced a new function: XLOOKUP. This attempts to address this shortcoming with, in my opinion, limited success. I discuss XLOOKUP further in Chapter 7.

really large spreadsheets with many calculations because it can take a long time to recalculate all of the affected cells. This can significantly slow down the development process.

To get around this, Excel provides the option to recalculate the values in the cells only when the user requests they be updated. To do this select File > Options > Formulas and specify "Manual Recalculation". The risk with this approach is that any report will have obsolete data if a manual recalculation isn't requested before the report is presented (i.e. printed or viewed).

Just Plain Frustrations: Excel tries, it really does. As good as Excel is, there are still a number of things it does that are just plain frustrating:

- There are places in Excel where the developer is presented a window with options to be selected or data to be specified. Unfortunately, some of these windows can't be moved or expanded. Therefore, it becomes difficult to see and specify the desired option. In all fairness, recent versions of Excel have fixed some of these, but there are others that are still a work in process.

- There are also features that don't work quite as you would expect. One of the PivotTable options is "Preserve Cell Formatting on Refresh". However, when the PivotTable expands onto new rows it does not honor the formatting of the cells in the new rows. Instead, it uses the default cell formatting. You must manually reformat those cells.

- Excel makes assumptions about what you want to do, especially when entering numbers. If, for example, you have a column in a table containing a ratio and you enter 1:1, Excel assumes that it is a time (i.e. 1:01am) and the value in the cell becomes 0.042361. A similar problem occurs with dates. If you have a column containing fractions and you enter 13/7, Excel (correctly) treats it as text. On the other hand, if you enter 12/7, Excel interprets it as the date December 7, <current year>.

- Beware! When you are editing a range specification (for example, in a conditional format), you can't use the arrow keys to position the cursor. Excel assumes that this is part of the range specification and inserts the cell position into the range string. This can be a problem if the range specification has already grown too long to be entirely contained in the window (such as when an insert or delete expands the specification for a conditional format as described above).

- Keep smiling. Originally Excel only dealt with numbers and simple text strings. Since then it has grown to be a full function document generator, complete with a wide variety of word processing features. Like any good word processor, Excel provides a Spellcheck feature. But don't expect Excel to catch all your fat-finger-fups:
    - Spellcheck is not automatic; it must be invoked manually in the review tab.
    - If a single cell is selected, Excel will ask if you want to check the entire worksheet. If a range is selected, Excel only checks that range.
    - Acknowledging the fact that much of what appears on a spreadsheet is supposed to be meaningless (from a proper language perspective), Excel only flags as errors words that it thinks *might* be real words that are misspelled. Some examples:
      "This isnt a test" ← Identifies the missing quote

"xyz" ← Changes to "xyzzy" (huh?)
"XYZ" ← No change (ignore all caps)
"zzz" ← Changes to "zazen" (maybe?)
"zzzz" ← No change (why?)

- And my Favorite: You open a spreadsheet and suddenly a message box like this pops up –

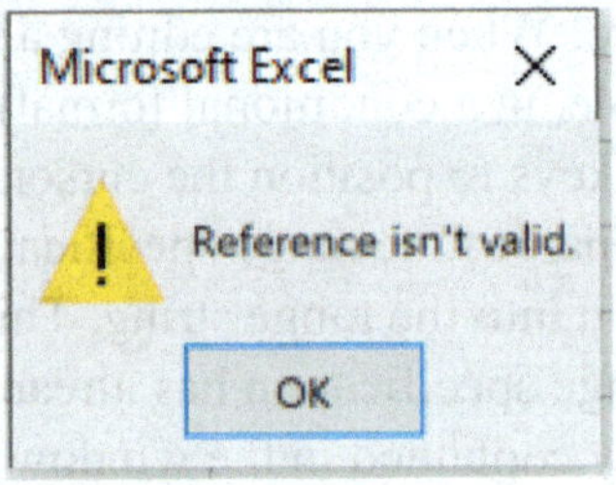

OK, What does this mean? What reference isn't valid? Where is the errant reference located? Excel won't tell you.
This is just one example. There are several situations where this type of thing happens. Microsoft has a long history of providing its customers with cryptic and less-than-useful error messages. Sadly, the tradition continues. I have found the best thing to do is a web search. There are hundreds of helpful web sites, blogs, and videos that can usually inform you as to what the message means and/or why the error occurred. Finding and fixing it is up to you.

5. THE BIG PICTURE

There is an old proverb:

Cleanliness is next to Godliness.

As computer systems have evolved and grown in complexity the practitioners of information technology have had to live with a updated version:

When it comes to systems,
Cleanliness is next to Impossible.

The operative word here is *evolved*. Most computer systems start out to perform a few simple tasks. The users of these systems generally like what they see, and then get ideas for how they can be enhanced and expanded. Soon there is an endless stream of upgrade requests, which lead to backlogs, methodologies, and steering committees. The developers, being the supportive people they are and sensitive to the needs of their users (and cognizant of the influence they might have on the next round of pay raises) do everything they can to jam as much functionality into the programs as quickly as possible. This usually means taking shortcuts wherever possible. The result is that these once elegant systems often *evolve* (there's that word again!) into beasts that are large, unwieldly and barely supportable.

Evolution, be it in nature or computer systems, is almost always a messy process. But while nature does a very effective job pruning the branches of the evolutionary tree that are no longer viable, computer systems aren't so lucky.

Often by the time the beasts are recognized as such they are so firmly embedded in the organization that there is no choice but to continue feeding and watering them.

The earliest general purpose digital computers were invented in the 1940s. Do the math and you see that what we now refer to as information technology has been around for over 70 years. During that time computers have grown from laboratory curiosities to being the central engine driving organizational operations and decision making. Along the way the technology organizations learned many important lessons about how to build and maintain systems so they don't become unmanageable beasts. Among the lessons learned are that the delivery of good, effective and maintainable systems depend on the adherence of structure, organization and standards.

Structure, Organization, Standards

The problem was (and is) that the users perceive these activities as non-productive overhead that doesn't contribute to the delivery of the product. In their defense, there is some justification for this feeling. It is precisely to address this issue that so much focus has been placed on "Agile" and related approaches to software delivery.

One of the reasons that PC-based productivity tools such as electronic spreadsheet software were embraced so quickly was that there was (and to a degree still is) the belief on the part of the users that they could eliminate this unnecessary overhead. There were several apparently good reasons for this belief:

- First, since the development process was now entirely in the hands of each individual user they could set priorities without the need to compete with other organizations for scarce resources.
- Second, since now the customer (user) and developer were the same person there are no translation issues. The developer/user implicitly understood the requirements and was in the best position to implement the solution.
- Third, the intuitive nature of the spreadsheet software meant that the messy evolutionary forces that had plagued the development of traditional computer systems were no longer a consideration.

Unfortunately, this wasn't the case. With respect to the perils of evolution, spreadsheets are no different than the procedural language computer systems that preceded them. Most spreadsheets start out simple – often a single table and a few simple formulas. Soon additional tables are added and the formulas become more complex. It isn't long before the spreadsheet has become, like the systems before it, large, unwieldly and barely supportable.

Now for the good news: The developers of traditional computer systems using procedural programming languages learned, after many painful years, that structure, organization and standards are the key to developing sustainable systems. Spreadsheet developers can take advantage of these lessons learned (if they are so inclined) without the necessity of experiencing the same trauma. But be warned: without structure, organization and standards the inevitable result will be the same.

As discussed in chapter one, a spreadsheet is a tool for solving a business problem by transforming some available data into the desired information. The desired information is the output. This can in the form of a report, a data file to be used by another system (or as input to another spreadsheet) or accessed directly via a link. The transformation consists of the arrangement of the various formulas, functions and features that act on data loaded into various tables.

Generically, every spreadsheet has three components:

- **Input:** There must be some mechanism for entering data into the spreadsheet.
- **Transformation:** Formulas that use the input data and change it or add to it to create useful information.
- **Presentation:** Some mechanism for taking the transformed data and making it available to the user. Note that the user can be a person, or another computer application.

Depending on the complexity of the problem, these three components could be implemented as multiple worksheets within a workbook, multiple tables, or even simply different columns within a single table. Regardless of how they are implemented, conceptually these three elements always exist. But – and this is important – how they are implemented will have a major impact on how easily the spreadsheet can be developed, used, and maintained.

Figure 6 on the next page shows the main components of a generalized complex spreadsheet.

The dotted box at the top represents various sources for the data that will be input into the spreadsheet. These sources could be computer files created by some other system, paper documents, or some other external source. The dotted box at the bottom represents the various places where the information from the spreadsheet could used. The contents of a spreadsheet table could be downloaded to create a computer file that would then be used as input to some other system. They could be printed as a report. Some external system could link to the spreadsheet and access the contents of a table directly.

The solid box in the middle represents the generalized spreadsheet. The spreadsheet can have one or more tabs, and each tab can be formatted with one or more tables.[24]

---

[24] As a general rule, it is a good idea to only have one major table in each tab. This makes the tables much easier to maintain and the logic much easier to follow. Most of the examples that I present in this book are structured that way.

FIGURE 6. GENERALIZED SPREADSHEET SCHEMATIC

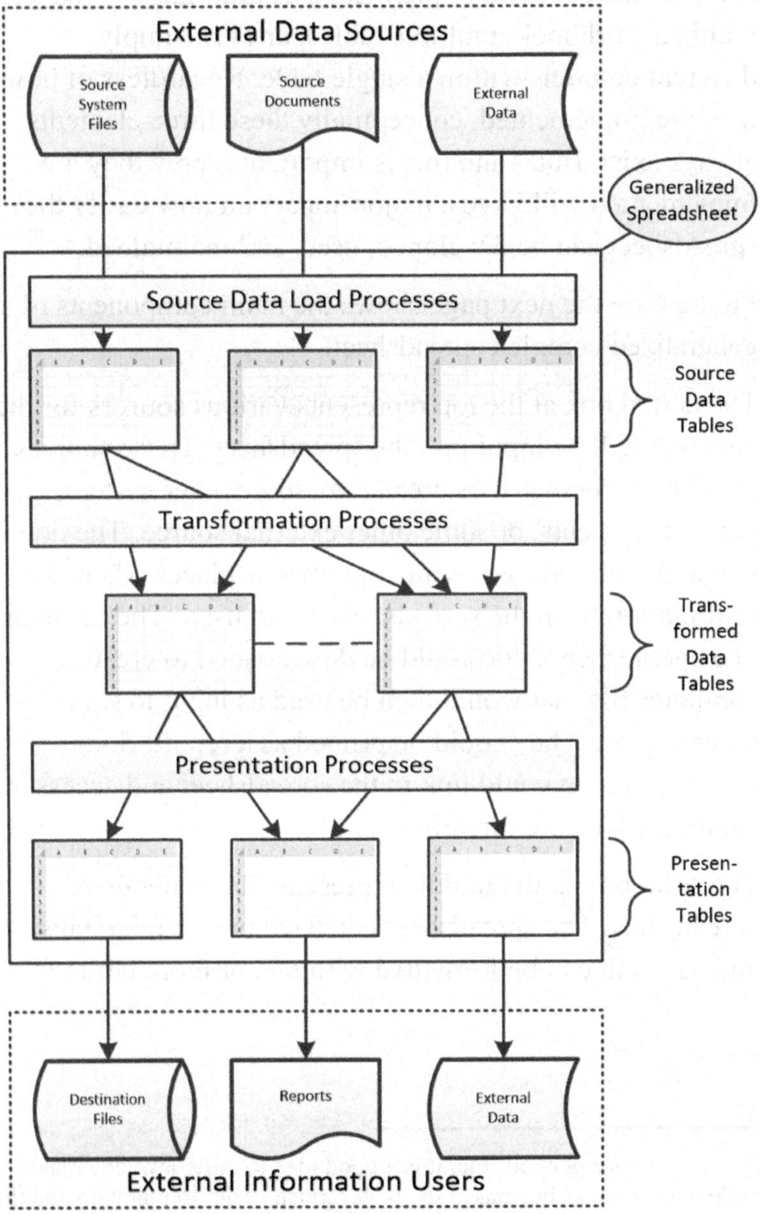

Given the generalized structure described above, and the fact that table is the unit of storage in a spreadsheet, it is not surprising that the tables in a spreadsheet fall into three groups:

- **Source Data Tables** – These are the tables where the source data are loaded. Once loaded, the contents of these tables are generally unmodified and are the same as the external source data.

- **Transformation Tables** – These are the tables where the data is "cleaned", filtered, combined and otherwise modified to support the final solution.

- **Presentation Tables** – These tables use the data from the Transformation tables and organizes it into the format required by the ultimate external user.

The processes that precede each set of tables consist of the various formulas, validation rules, functions, selection criteria and activities that process the data as it moves from one set of tables to the next. Depending on the structure and organization of any specific spreadsheet these processes could be manual or completely automatic. These three groups of processes are the subject of the following chapters.

6. INPUT: THE PRIMACY OF DATA

The heart of every spreadsheet is the data. Without data, a spreadsheet is meaningless. Without the ability to change the data or enter new data, a spreadsheet is only good for a single use.

The first step in building a spreadsheet is understanding the data content of the problem. To do this the developer must answer three questions specifically related to the data:

- What information is needed to produce the solution?
- What data is available as input to the transformation process?
- How can the data best be organized into one or more tables so that it can be transformed?

This last point is critical. The basic data structure of a spreadsheet is the table. As discussed earlier, a table can be defined as a data structure consisting of columns (each of which represents a unique data element) and rows (each of which represents a unique occurrence of those data elements).[25] Most of the functionality provided by

[25] It is possible to build tables that are organized sideways – the rows are the individual elements and the columns are the occurrences. As a general rule Excel and other spreadsheet programs do not handle this very well. There are a (very) few situations where it is

spreadsheet software is designed to work efficiently and effectively on data that is organized in tables.

Once the available data and how it relates to the solution are understood, the data must somehow be "loaded" into the spreadsheet. While loading data into a spreadsheet can properly be considered an aspect of using the spreadsheet (different data might be loaded many different times or by many different users), the mechanisms for loading the data must be built as part of the development activities.

In this chapter I will first discuss how to recognize the data content of a problem and analyze how that data can be organized into usable tables. Then I will discuss various methods for loading the raw data into those usable tables.

---

appropriate to do this. I will give some examples later in the book.

Evidence

A police detective once told me that evidence is the data content of an event. In the situations he dealt with the event is a crime of some kind that has been committed. The evidence collected is data that is (hopefully) related to the crime. His challenge was to collect as much of the data as possible and determine if and how that data relates to (and therefor be useful in solving) the crime.

In many respects our challenge is very similar. Instead of a crime we are presented with a problem, and we need to determine what data we have and how it can be used to solve it. Sometimes the available data and its relationship to the solution are obvious. Sometimes, not so much.

FORM & FUNCTION

Financial Applications are a good example of problems where the data content is usually quite obvious. (Though it wasn't generally understood at the time, the simplicity of their data structures is one reason that financial applications were among the first to be successfully automated). The various systems which process financial information (Accounts Payable, Accounts Receivable, Fixed Assets, Loan Processing, Payroll, General Ledger, etc.) each create files[26] with their own unique format (i.e. collection of data

[26] I am using the terms files (and records) here very generically. They could be in any form , from a physical file on a sequentially accessed media to entries in a database. Their form is not

elements relevant to that specific application). At the most basic level, these files come in two general types:

- **Transaction Files:** Files that contain multiple occurrences of records, each of which contains information about a specific event and has an explicit set of data elements. For example, the Accounts Payable system may produce a file with a record for each payment made. The elements in the file could include Payment ID, Payee ID, Amount, Date, PO Number, etc.

- **Reference Files:** Files that are relatively static and contain information that supplements, constrains or clarifies specific elements in the transaction files. In the Accounts Payable example, there could be a Payee file. The elements in the file could include Payee ID, name, address, phone, bank account information, payment terms, etc.

It is typically an easy task to map both of these file types into a spreadsheet table structure.

Since each of the applications that produce financial information has its own unique collection of data, it makes sense that the data for each file would be loaded into its own table with its own unique structure (i.e. column definitions). Each row of the table would represent a single transaction (in the case of transaction files) or entity (in the case of reference files).

There are several important points that need to be considered when building these tables:

---

what's important.

- A well-designed spreadsheet should be able to function for a wide range of the data. Depending on the application, you wouldn't want to create a separate spreadsheet for each month of the year, or for each customer, or whatever. Instead, it might be possible to create monthly, quarterly, year-to-date, etc. reports from the same spreadsheet!

- For transaction files, the number of *rows of interest* would depend on the scope of the events under consideration. For example, if the spreadsheet is to present information about the activity for a fiscal period, one of the selection criteria would include the range of dates under consideration.
  By contrast, reference files will often be loaded in their entirety. Using the Accounts Payable example, the entire payee reference file would be loaded because which payees would be needed would vary depending on the specific options and selection criteria for the particular report being produced.

- To be useful, it must be possible to *link* the elements in the transaction file(s) to the corresponding reference file(s). This requires a *key value* that is common to both. For example, the payment file contains a Payee ID, and each entry in the payee reference file is identified by their Payee ID. As discussed in Chapter 4, it is critical that these fields have identical form and format. Otherwise, It can be significantly more difficult (or impossible) to tie the two together.

- Wherever possible, avoid redundancy. Even if a transaction file has some of the same information as the corresponding reference file, use the data from the reference file. The reason for this is simple: any redundancy is an invitation for inconsistency! Only use data from a transaction file that specifically relates to the transaction and the problem being addressed. Get everything else from the appropriate reference files.

So far so good. We've considered a class of applications where the data content and structure are pretty obvious and well defined. But what if they aren't? Consider the following request:

Create a spreadsheet model to optimize the flow of traffic through a section of highway in a downtown metropolitan area.

Spreadsheets are great tools for building models. But how should the tables be organized? What are the events?

The process is the same, but there are some extra steps to getting there:

- What (in explicit, actionable terms) is the desired result? For the traffic model, what does "optimize" mean? Does it mean maximize the number of cars moving through the section of highway under consideration, or does It mean minimize the number of accidents? Related questions that need to be answered might include:
    - What is the exact definition of the section of highway (from where to where)?
    - What is the timeframe of interest (24 hours or over an entire year)?
    - Is the configuration of the highway fixed or can it change during the day? (This is exactly what traffic engineers evaluate when configuring express lanes.)

- What data is available or can be collected? Are there current traffic studies that can be used? If so, what format are they in?

- What is the appropriate *unit of measure*?

UNIT OF MEASURE

This last question is central to solving the problem and must be answered correctly to build an effective solution. Without the correct definition of the unit of measure you can't identify the relevant events. A relevant event is (by definition) an activity that has an impact on the unit of measure; if it doesn't, the event is irrelevant and can (or should) be excluded. Without knowing the events you can't determine what data is needed.

The first part of the answer to the unit of measure question is to define "what" the unit is a measure of. In the financial applications example the "what" is usually *money*. (That's why they're called financial applications - duh!) Each event represents something that has a monetary impact. For the traffic example, the unit of measure is probably *time interval*. The clue that leads to that conclusion is the requirement to optimize *flow* (which can be defined as some amount of something per unit of time). From this it logically follows that the events are vehicles passing through a specific point each time interval.

The second part of the answer to the unit of measure question is to define "how much" each unit represents. In the financial applications example the how much is usually defined by the currency being used. In the United States we deal with dollars. Depending on the application, this often includes cents (i.e. dollars and two decimal places). Budgeting for a large corporation or government agency might be entirely in $000s. In other areas this can be different[27]. For the traffic example the answer is not so obvious. Time intervals can be measured anywhere from microseconds to millennia.

The choice of interval can have a significant impact on the value of the data collected, and the effort required to obtain it. If the interval is too small the number of data points becomes excessively large, with many of them being meaningless. If the interval is too large important details may be lost in gross averages.

To demonstrate this, let's look again at the traffic flow problem. Before the problem can be analyzed with a spreadsheet, data about the current traffic situation must be collected and analyzed. You've probably seen those hoses strung across the road that the cars drive over. Each time a car or truck drives over the hose it registers a count. Counts are totaled over the specified interval. This total is compared to the "capacity" of the road (i.e. maximum number of vehicles the road can handle).

Consider the chart (below).

---

[27] I once worked on a payroll application for the native workers at an African rubber plantation who were paid in pounds of rice!

FIGURE 7 TRAFFIC FLOW SAMPLE ANALYSIS

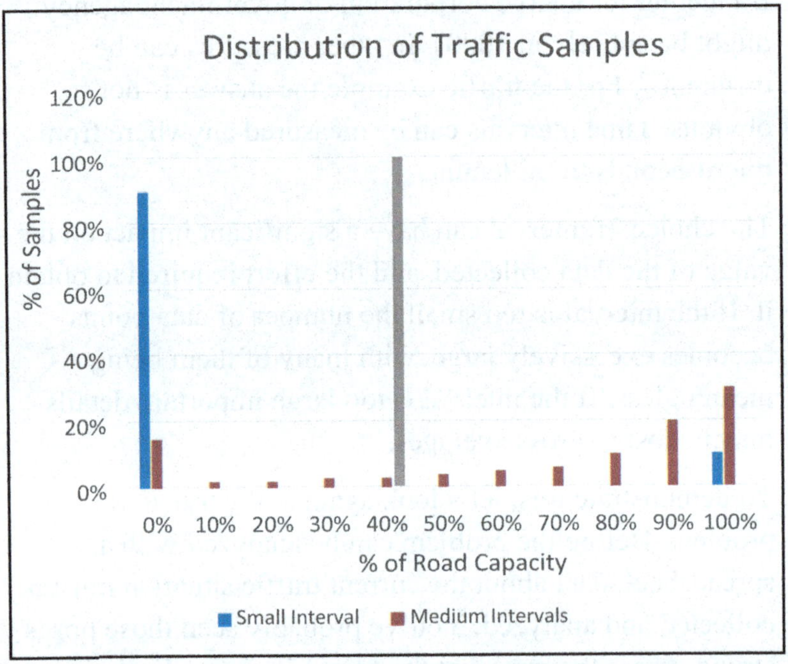

It was prepared from a model built to validate traffic data that was being collected. The chart shows three different results based on the choice of time interval: If the sample interval is very small (for example, 0.1 second) you see that about 90% of the intervals show no activity at all, and 10% show the road at 100% capacity. Based on this graph, you might assume that the road has no capacity issues since 90% of the time the road is completely unused.

If the sample interval is very large (say, 24 hours) everything gets smoothed into an average of about 40%. Again, this might indicate that there are no capacity issues.

If the sample interval is somewhere in the middle (10 minutes) a completely different, and more realistic, picture emerges. There are periods when the road is used very little or not at all. Not surprising for the middle of the night. But there are also periods when the utilization becomes dangerously high. When you examine the time-stamp for each of the samples it's not surprising those high percentages correspond to rush hour.

The lesson here: choice of unit of measure (what and how much) will have a material impact on the quality of the data being collected. Spreadsheets can't do magic: without good data, there's little a spreadsheet can do.

This consideration actually cuts both ways. In some situations (such as the traffic problem described above) you may have some control over the data being collected. In these situations you need to consider the characteristics of the final solution and work backwards to determine the characteristics of the data needed to support it. (Note that I didn't say work backwards to determine what data you need to support your pre-specified desired solution. That would be cheating!) In other situations (such as the financial applications) you may be stuck with what they are able to provide. This in turn may constrain the types of solutions you can produce.

There is one more cardinal rule about data that you must consider when deciding how to approach a problem: Aggregating is easy; dis-aggregating is hard, or impossible. It's easy to take a bunch of data points and add them together to get a total. It can be functionally impossible to take a total and determine what it consists of[28].

Spreadsheets should be designed to accept the most detailed data available. This provides the greatest flexibility when transforming the data into useful information.

ORGANIZATION

As mentioned previously, the data in spreadsheets are organized into tables. Tables consist of rows and columns. Each row represents an event; each column represents a unique piece of data *specific to that type of event*. Each cell, the intersection of a row and a column, contains a unique piece of data about a specific event. This is the organization required to support the transformation process – converting raw data into useful information.

It should go without saying, but it is worth repeating: Each table that contains raw (input) data should only contain the data for one type of event. If the problem being addressed deals with six types of events, there should be six separate tables to accept that data.

Sadly, the world doesn't always cooperate. Whether the data being inputted is already available in table form or not, the developer must somehow get it there. Sometimes this can be a multi-step process.

The process of designing a table begins with two questions:

---

[28] This *mathematical asymmetry* is the basis of the *public key cryptosystems*, which are what enables us to do all those things we do on the internet.

- ➢ What types of data are present in each event?
- ➢ What is it that makes each event unique?

The answer to the first question becomes the definition of the columns. The answer to the second question identifies the rows. Often it is the second question that is the hardest to answer.

If the data is coming from another system there is often a unique identifier already associated with each event – Transaction files usually have some type of transaction ID created by the originating system. Reference files usually have a system assigned identifier that uniquely identifies each entity. If the data is not coming from another system the user may have to create a unique identifier as part of the data input process.

Spreadsheets have no problem with tables that have multiple identical rows, so why bother? In a well-designed spreadsheet it should always be possible to verify that each entry in the tables exactly match the corresponding source data. This is essential for verification and for auditing. Wherever possible, spreadsheet tables should include sufficient data elements to make each row unique. As a last resort (and also as just a good general practice), the data should contain a time-stamp, as it is unlikely that two events of the same type can occur at exactly the same moment.

In the traffic flow problem a time stamp is about the only identifying data we have. The rubber hose lying on the street doesn't record license plates. The only data collected is the location of the counter and the number of cars that pass every 10 minutes. By knowing the time of each count, the data can be linked to weather, police, fire and other data which can directly impact the traffic flow.

The first step in designing and developing an effective spreadsheet is to understand the available data, the events that impact that data, and how the data relates to the desired information.

Many Paths, One Destination

In the 1939 movie *The Wizard of Oz* there is a scene where Dorothy (Judy Garland) comes to a fork in the yellow brick road. She is unsure as to which is the right path to take. It is there that she meets the scarecrow, who explains to her that a problem can have more than one solution. The rest, as they say, is history. Armed with this new insight, she proceeds to recruit unlikely allies, push through intractable bureaucracies, defeat overwhelmingly powerful adversaries, and become an early (though unappreciated) pioneer of space travel.

Once the data content of the problem has been assessed, the next challenge is to devise mechanisms for getting the source data into the spreadsheet. As with Dorothy's dilemma, there are often several viable approaches. Determining which is best will depend on several factors:

- **How many different sources are needed to provide all the required data?**
    - How many different transaction files?
    - How many different reference files?
    - What types of "control data" are required? For example, if the spreadsheet will produce reports for multiple time-periods or different groupings of entities (i.e., customers), there must be a way to specify this selection criteria.

- **What are the characteristics of the source data?**
    - Does each type of source data have a single format, or are there multiple formats?
    - Are the formats fixed or variable?

- Are the individual data fields presented separately or are multiple data fields imbedded in a single source field?
- Is there a unique key value associated with each record?
- What is the maximum number of records that will be loaded each time the spreadsheet is updates?

➢ **What types of media contain the required source data?**
- Paper forms?
- Sequential computer files?
- Records in a relational database?
- Are they local or accessed over a network?
- Are they accessible by the PC running the spreadsheet software?

It is important to keep in mind that the objective is to get the source data, however it is initially formatted and organized, into one or more well-defined and correctly structured tables in the spreadsheet. This may require multiple steps. The first step in each case is to get the source data into the *Main Spreadsheet*. A useful convention is to load the source data as is into a "Raw Data" table. The cells in the "Raw Data" table can then be referenced by another "Fixed Data" table. The cells in the "Fixed Data" table can reference specific columns (or portions of columns – more on this later) in the "Raw Data" table and apply appropriate edits or transformations to produce a correctly structured table with clean data.

STEP 1: SOURCE MEDIA TO RAW DATA TABLE

There are four basic ways to get data into a spreadsheet:

- **Keying the data (Manual Entry).** If the data to be loaded into the main spreadsheet is being provided on a paper form (or equivalent), manual entry may be the best (or only) option. Sometimes it is possible to scan a form and store it in a format that can be imported into Excel. If this is possible then this is the preferred approach, as it reduces the likelihood of entry errors. If not, it will have to be keyed.

  The columns of the Raw Data table into which the form will be keyed should be laid out exactly as the fields appear on the form. Any rearrangement of the columns needed to support the further processing of the data can (and should) be done in the Fixed Data table. The person keying the data should not have to skip around either the form or the table. There should be header rows frozen at the top of the table that match those on the form, and the fields (columns) should be formatted to match the fields on the form (in particular dates and numbers).

  When designing a spreadsheet that will require keying data from a form there is a critical question that must be addressed: Will the data on the form be accepted and entered as is, regardless of whether there is missing or incorrect data, or will errors be flagged and corrected at entry? This is a critical question, as it will affect the entire design of the initial data entry process. If the person doing the data entry is in close proximity to the data source, it might be better to correct the data as it is entered as this will facilitate all the processing of that data going forward. In this situation, data validation criteria might be added to specific fields where possible. If not, then it might be necessary to accept whatever is provided, identify the mistakes, flag them somehow, and provide a means for correcting them further "downstream"[29].

- **Copying the data.** If the source data to be loaded into the main spreadsheet already exists in a computer file it can often be copied directly into the Raw Data table. This is accomplished by opening the file with either Excel or the applicable application based on the file type, selecting (highlighting) the required data on the screen, and using the Windows copy / paste functions to copy the data into the Raw Data table in the main spreadsheet.

  It is also possible to copy / paste data directly from some web screens, but this usually requires some additional adjustment before the data can be used. This process is discussed in detail in the Special Situations section at the end of this chapter.

  Regardless of where the data originates, what is most important is that the data be in consistent rows and columns when it is copied into the Raw Data table. If the source data file is being opened by Excel it will already be organized this way, which makes the copying process much simpler. If not, it will be necessary to adjust the data before it is copied into the Raw Data table, or as additional steps in the cleanup process from raw data to fixed data.

  Many systems produce files in an Excel compatible format. Certain file types (such as .csv – comma separated variables) can be opened directly by Excel. Excel is the default application for processing .csv files.

---

[29] A good example of this: One project I worked on involved field agents from across the country faxing in forms that were keyed into a spreadsheet. The forms frequently had missing or incorrect data. The person entering the forms keyed in what was supplied, and the spreadsheet would produce a "bad data" report. A specialist would follow-up with the agents (sometimes days later) to correct the issues. The transactions the forms represented were kept in a hold status until the errors were corrected.

When a user opens a .csv file (double-click on it or right-click then select OPEN) Windows will launch Excel and pass the file name to it for processing.
Excel can be used to open many other file types. Launch Excel directly from the START menu and select blank spreadsheet. Invoke File > Open and specify the name of the file to be loaded. If the format is not one that it recognizes Excel will launch the Text Import Wizard. The wizard will ask questions about the data in the file, including where and how the columns should be separated. This can be very helpful if, for example, a report is to be imported.
Note: In each of these cases there will be two Excel spreadsheets open – the *Main Spreadsheet* (which will have the Raw Data table(s), the Fixed Data table(s), and the other tables necessary to produce the final reports or other information) and a *Source Data Spreadsheet* (which will have some type of Source Data Table where the source data is initially loaded). Once the data is copied from the source data spreadsheet into the Raw Data table the Source Data Spreadsheet can be closed and deleted or archived.
This process is shown schematically in the following diagram:

FIGURE 8. IMPORTING SOURCE DATA

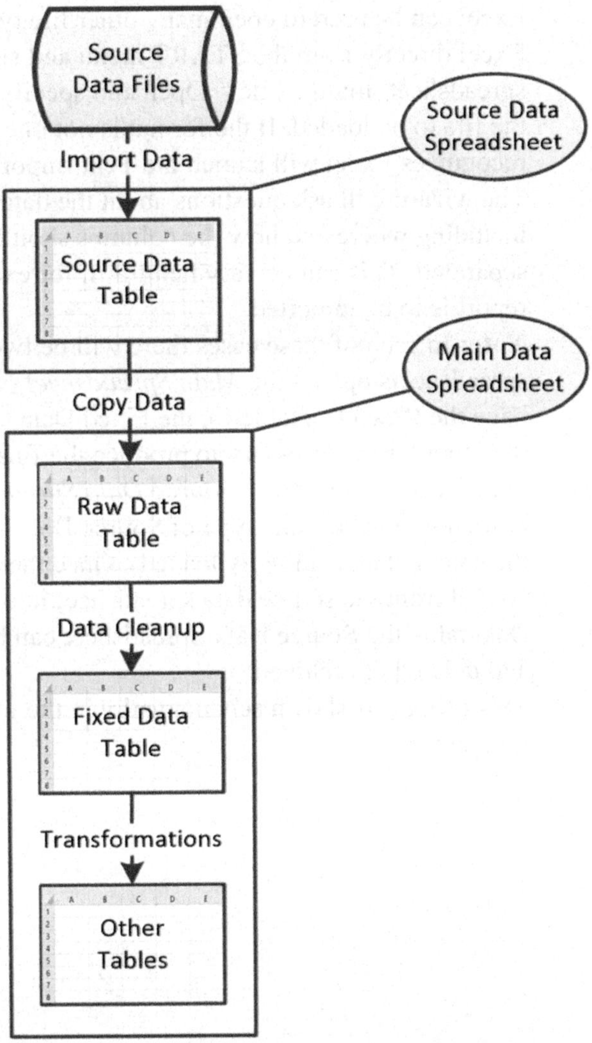

➤ **Moving the data.** Sometimes, for a variety of reasons, copying the data form the source data spreadsheet into the Raw Data table may not be feasible. Consider for example a report is being imported as described above.

The report could be several hundred pages, which would make selecting the data and copying it tedious at best and possibly infeasible. A better option would be to rename the tab in the source data spreadsheet "Raw Data", and then *move* the tab into the main spreadsheet. But here's the wrinkle: Formulas in the Fixed Data table can't directly reference any cells in the new Raw Data tab. If the Raw Data table initially exists and is subsequently deleted to bring in a new Source Data Table, any formulas referring to cells in the Raw Data tab will now get a #Ref error.

Excel provides an elegant way around this problem – the INDIRECT function. The value of the cell referenced by most functions is used as a parameter to that function. The value of a cell referenced by the INDIRECT function is interpreted as the name or label of the cell whose value is to be used. This can be seen in the following example:

In this spreadsheet there is a tab named "Raw Data". Cell A1 of the Raw Data tab contains the text string "Source Data".

FIGURE 9. USING INDIRECT TO IMPORT DATA

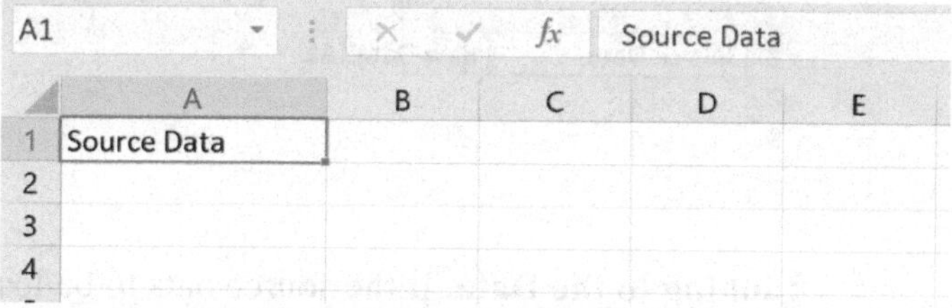

Another tab in the spreadsheet ("Fixed Data"), has a formula in cell A1 that references cell A1 in the Raw Data tab.

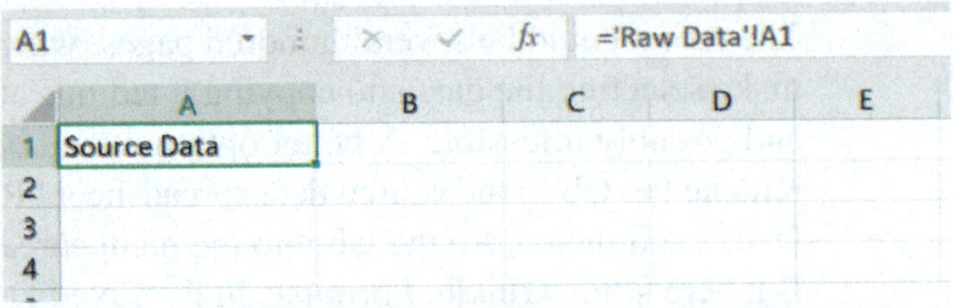

If the Raw Data tab is deleted, the formula reference is lost, and will now return a #REF error (even if the Raw Data tab is subsequently restored).

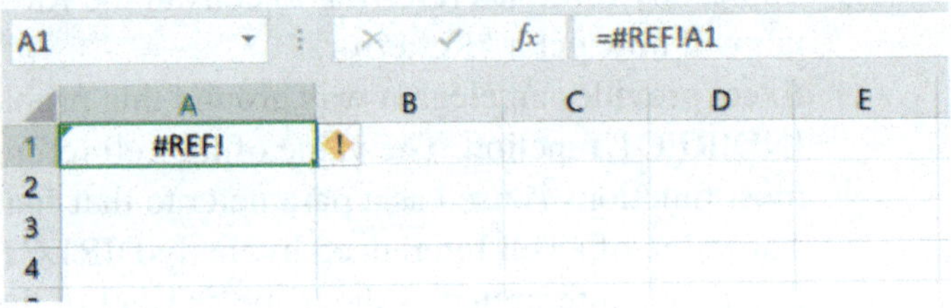

The beauty of the INDIRECT function is that it is only evaluated when the cell containing the function is accessed. If at that time the Raw Data tab exists then it will return the correct value. If the Raw Data tab doesn't exist it will return an error.

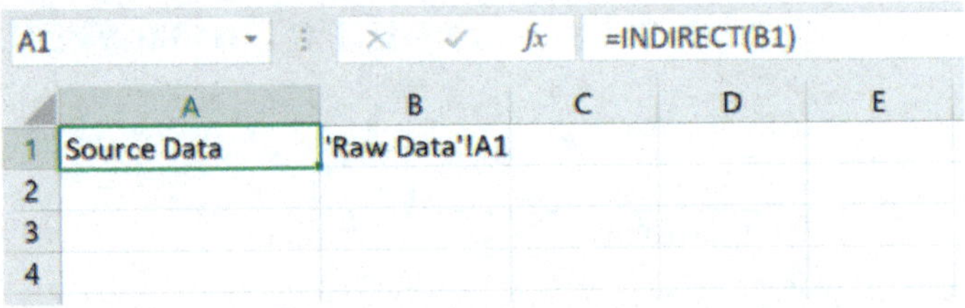

➢ **Linking to the Data.** If the source data to be loaded into the main spreadsheet already exists in another spreadsheet (or some other appropriately formatted relational table) it can be linked to directly. This can save the effort of loading the data using one of the methods described above. The downside to this approach is that it is very dependent on the accessibility and naming conventions of the file and environment where the source data file is located.

When the main spreadsheet is opened it will attempt to resolve any links to external data sources. If it cannot locate the specified file it will produce an error message and ask the user to either edit the link or use the data from the last time the spreadsheet was opened. This can be a major concern if the spreadsheet is intended to be portable – that is, used by different users in different locations.

One more caveat: it is best to copy the linked data into the main spreadsheet as *values*. This way the data remains and the spreadsheet continues to function correctly even if the link to the source data is broken.

Regardless of how the source data is to be loaded into the main spreadsheet, there is one extremely important consideration that must be incorporated: The Raw Data table must be designed to handle more than the maximum number of rows that will ever be loaded on any single update of the spreadsheet. If not, at some point the data will overflow the Raw Data table and those rows won't be recognized by subsequent steps in the processing. As discussed previously, place a control row just below the Raw Data table and include a test to identify if an overflow has occurred.

STEP2: RAW DATA TABLE TO FIXED DATA TABLE

Once the source data has been loaded into the Raw Data table, it needs to be scrubbed and organized. Scrubbing refers to identifying and correcting any issues in the source data such as invalid or missing data elements. Organizing refers to arranging the validated data elements in the best way to support the requirements of subsequent steps in the transformation process.

Scrubbing: News Flash – It's a dirty world! Regardless of where the data originates or how it is imported into the Raw Data table of the spreadsheet it can contain errors. The main purpose of the Fixed Data table is to ensure that all the data is error-free before it is passed to any subsequent process or report. These errors typically fall into one of several general categories, and the tests to identify them should be performed in order:

- The first test that must be performed against each <u>row</u> in the Raw Data table is to determine if it contains valid data. Consider the situation where a report is being imported into the spreadsheet. The report probably has header and footer rows on each page. It may also have total rows or other breaks between the individual detail lines that contain the elements of interest.
  Even if the data was imported from another spreadsheet and already has cleanly delineated data elements, it is still important to identify the valid rows. The Raw Data tables and Fixed Data tables should always be set up with more rows than are ever expected to be imported from the source data. Therefore, there should always be empty rows at the bottom of the table that must be

flagged as not valid (and as discussed previously, a test to determine if the source data import overflowed the Raw Data table).

Determining what rows are valid will depend on the type of data being imported. Often this will be based on testing the characteristics of a specific cell (i.e., column). If the rows that contain the data of interest have a number (such as a date) in a specific column, that can be tested and a determination made. Sometimes it is necessary to test multiple cells (columns). Any cobination of tests can be performed. Consider the following hypothetical test which would be used to set a flag in the "Valid?" column in row 17 of the Fixed Data table.[30]

```
=IF(AND(ISNUMBER(RAW!G17),ISTEXT(RAW!C17),
RAW!J17>40000,NOT(ISBLANK(B17)),1,0)
```

The convention here, is that a 1 means the test passed (i.e., the row is valid) and 0 means it failed (i.e., the row is not valid). This result can then be used in all subsequent tests. In addition, a simple SUM of the Valid? Column will yield the number of valid data rows that have been imported.

As a general rule, it is better to have fewer tests than more. The reason is simple: A row might be valid (i.e. represent an actual transaction), but could also have missing or incorrect data. It is better to recognize this situation and flag it as an error than to exclude an otherwise valid row as not being of interest.

It goes without saying, but it is worth repeating: The

[30] As a matter of convenience, I always try to structure the Raw Data table(s) and the corresponding Fixed Data table (s) so that the rows are in alignment. In other words, row 17 in the Fixed Data table is referencing data in row 17 of the Raw Data table, etc. If necessary I create an appropriate number of extra header rows on each table.

spreadsheet developer must understand the data to create an accurate and complete set of validity tests.

- Once the valid rows have been identified, it is necessary to extract each of the required data elements. If the source data was already in neatly defined columns then extracting won't be necessary. Unfortunately that often isn't the case.
  Consider the Source Data Table shown below. This is part of a download from a bank web site. Financial web sites (and many others) often provide a download capability that will automatically extract some portion of the activity and store it in a .csv file. This is how the downloaded file looks when it is opened using Excel:

	A	B	C	D	E
1	Date	Description	Withdrawals	Deposits	Balance
2	9/9/2019	CHECK 4974 ACME WIDGETS	$614.66		$49,014.39
3	9/9/2019	PAYMENT,E-CHECK 4971 GEORGETOWN ELECTR CHECK PYMT	$312.48		$49,629.05
4	9/6/2019	CHECK 4970 WASTE DISPOSAL SERVICES	$3,261.44		$49,941.53
5	8/31/2019	ACCTS RECBL CUSTOMER NUMBER 21673TFW		$245.80	$53,202.97
6	8/31/2019	ACCTS RECBL CUSTOMER NUMBER 46892		$2,934.80	$52,957.17
7	8/25/2019	CHECK 4969 INDUSTRIAL CLEANING	$165.00		$50,022.37
8	8/24/2019	PAYMENT,E-CHECK 4968 GOOGLE CHECK PYMT	$126.19		$50,187.37
9	8/23/2019	CHECK 4962 BRUCE'S HARDWARE	$50.00		$50,313.56
10	8/22/2019	INTEREST PAYMENT		$41.17	$50,363.56
11	8/22/2019	PAYMENT,E-CHECK 4964 CHASE STATIONARY CHECK PYMT	$500.00		$50,322.39
12	8/19/2019	ACH DEBIT 2918 FRIENDLY FINANCE	$2,336.31		$50,822.39
13	8/19/2019	CASH DEPOSIT 4619VFW		419.26	$53,158.70

The Date, Withdrawals, Deposit and Balance amounts are already neatly loaded into separate columns. The Description column appears to have multiple interesting fields imbedded within it – Activity Type, Check Number, Vendor Name / Customer ID. The MID function can be used to extract parts of the description into separate cells. The challenge is how to determine where each field begins and ends.
Part of the problem is that most of the commonly used fonts are *proportional*. This means that the width of the character on the display or on paper varies. A "W" will be wider than an "1". Spaces are narrower still. This makes the display or print look better because the amount of space between the characters is constant. But it does make it hard to determine how many characters

are in each line (which must be known to set up the MID function).
This problem can be simplified by formatting the Raw Data table with a *fixed width* font such as Courier New. It's also useful to have a counter as part of the header to show exactly what column each element is in. Copy the data into the Raw Data table using the "paste special > values" option to preserve the formatting:

	A	B	C	D	F
1	Date	Description	Withdrawals	Deposits	Balance
2		0 1 2 3 4 5 6			
3		1234567890123456789012345678901234567890123456789012345678901234 5			
4	Date	Desc	Wd	Dep	Bal
5	9/9/2019	CHECK 4974 ACME WIDGETS	$614.66		$49,014.38
6	9/9/2019	PAYMENT,E-CHECK 4971 GEORGETOWN ELECTR CHECK PYMT	$312.48		$49,629.05
7	9/6/2019	CHECK 4970 WASTE DISPOSAL SERVICES	$3,261.44		$49,941.53
8	8/31/2019	ACCTS RECBL CUSTOMER NUMBER 21673CFW		$245.80	$53,202.97
9	8/31/2019	ACCTS RECBL CUSTOMER NUMBER 46892		$2,934.80	$52,957.17
10	8/25/2019	CHECK 4969 INDUSTRIAL CLEANING	$165.00		$50,022.37
11	8/24/2019	PAYMENT,E-CHECK 4960 GOOGLE CHECK PYMT	$120.19		$50,187.37
12	8/23/2019	CHECK 4962 BRUCK'S HARDWARE	$50.00		$50,313.56
13	8/22/2019	INTEREST PAYMENT		$41.17	$50,363.56
14	8/22/2019	PAYMENT,E-CHECK 4964 CHASE STATIONARY CHECK PYMT	$500.00		$50,322.39
15	8/19/2019	ACH DEBIT 2918 FRIENDLY FINANCE	$2,336.31		$50,822.39
16	8/19/2019	CASH DEPOSIT 4619VFW		419.26	$53,158.70
17					$52,739.44

The individual fields inside the Description are clearly identifiable and in fixed positions, which make them easy to extract. The scale at the top shows exactly what displacements ("start_num") and lengths ("num_chars") to use.
It should be noted that whenever text fields are being retrieved the field should be "trimmed" (using the TRIM function) to remove excess and trailing blanks. This ensures consistency, which can be critical in subsequent processing.

➢ Once the valid rows have been determined and the individual data elements identified, they should be tested. The test for each element will vary based on its expected meaning or purpose. Common tests include:
    - Required Fields – If a field is required, does it contain data? This can be determined using the ISBLANK function.
    - Data Type – Is the type of data in the field as expected (i.e. numeric or text)? This can be

determined using the ISNUMBER or ISTEXT functions.

- Data Range – This is most often an issue for numeric fields. For example, the field might contain a code that is supposed to be within a specific range. Or it might be a dollar amount that must be positive.
  If the data element is supposed to be a date it can be checked against an allowable range of dates. The minimum and maximum allowed dates can be put into *control fields* anywhere convenient in the spreadsheet[31]. Since dates and times are numeric values, it is easy to test if the field in the Raw Data table falls within the allowed range.
- Data Value – If the field has some special significance then it can be validated against the corresponding Reference table. For example, if column "E" of the Raw Data table is supposed to have the Customer ID, the Fixed Data table could contain the following in the row validating row 17 of the Raw Data Table:

```
=IF(ISNA(VLOOKUP(Raw!E17,Cust_Tbl,1,0)),1,0)
```

  In this example, the field that is supposed to contain a Customer ID in the Raw Data table is used to lookup the record for that customer in the Customer Reference Table (Cust_Tbl). If the Customer ID is not valid, the VLOOKUP will return "#N/A", the ISNA function will return "TRUE", and the field will be set to a 1. If the Customer ID is valid, the field will be set to 0. As above with the Valid? test, a SUM of this

---

[31] This will be discussed in detail in Chapter 9.

column will indicate exactly how many Customer ID errors have occurred.
  - Auto-filter can then be used to display only the rows which contain errors.

- Depending on where the source data originates and how it is loaded there may be overlaps from one run to the next, or from one data source to the next. Whether testing for duplicates is done in the Fixed Data table or further in the processing will depend on the specific application.
  Regardless of where the test for duplicates is done, the process is generally the same:
  - The data where there might be overlaps is sorted on some identifying value or combination of values that should be unique.[32]
  - A test is performed to determine if consecutive rows have the same identifying information. If they do, both rows are flagged as potential duplicates (the column labeled "Dup?" is set to 1). A SUM of the "Dup?" column will indicate exactly how many potential duplicates there are.
  - Auto-filter can then be used to display only the duplicated rows. Each pair of rows can be examined to determine which should be eliminated.

Organizing: Once all errors in the source data have been addressed, the corresponding Fixed Data table should be in a form that makes the data easy to use by all subsequent

---

[32] It goes without saying, but I'll say it anyway: The fields being used as the sort key should contain values, not formulas. If not, the results can be unexpected as the formulas are reevaluated based on their new position in the table.

processes in the spreadsheet. The Fixed Data table is not a mirror of the Raw Data table. Even where the rows are the same the columns can be completely different. In a very real sense, the Fixed Data table is the gatekeeper of the data.

How easy the Fixed Data table is to use obviously depends on many factors and will always be application specific. There are however several *general guidelines* that should be considered:

- First and foremost, every column in the table must have a single unique meaning and purpose. This should be obvious, as it is an intrinsic part of the definition of a well structured table. Unfortunately, many times people try to cheat for some mis-guided belief that it will improve efficiency. In reality, they are only creating problems and a lot more work for themselves later. There should also be a header row at the top of the table that clearly identifies the meaning of each column. The headers should be frozen so that they remain visible as one scrolls through the table. The headers should also be unique, so that the table can be used as the source data range for a PivotTable if desired.

- Directly related to this, each test that is performed on the data (valid row, missing or invalid data elements, duplicate rows, etc.) should have its own flag in its own column. A useful convention is to set the flag cells to one if the test (whatever it is) succeeds, and a zero if the test fails. Note that the meaning of the one or zero will differ for positive and negative tests.
  For example:
  - If the row is valid, the flag in the "Valid?" column should be one (positive test).

    - If there is a test to identify rows where an amount field is greater than $10,000, the flag in the "Amt>10k?" column should be one if the record is valid and the amount is greater than $10,000 (positive test).
    - If the account number is a required field, the flag in the "Acct Missing?" column should be one if the record is valid and the account number is missing (negative test). In other words, this test is for a negative (i.e., "bad") condition. If the test succeeds, a "bad" situation has occurred, and this needs to be noted.

- At the top of the table (usually directly under the header row) include a SUM for each column that contains a test, or that has some other numeric value that might be of significance. This row of sums should also be included in the frozen headers.
  For example, if the source data is being imported from a report, the SUM of the "Valid?" column will indicate how many records were recognized as valid. Or a SUM of an amount column could give a grand total for the table. These totals can be compared to control totals in the original report to ensure that all the data has been successfully loaded. The totals of the various test columns will indicate how many errors must be corrected before the data can be further processed.

- There's no rule that says the Fixed Data table can only contain data extracted from the Raw Data table. If there are data in other tables that are related and relevant, that data can also be pulled into the Fixed Data table.
  For example, if one of the fields in the Fixed Data table is a Customer ID, and customer information will be needed later, why not retrieve it from the Customer Reference table and include it as part of the Fixed Data

table.

- Arrange the columns in the Fixed Data table in an order that will facilitate subsequent processing. In particular, if each row of the table has a significant and unique "key" value, that field should be in the first column. This is especially important for reference tables so that they can easily be used in formulas containing the VLOOKUP function.

Special Situations

The road from source data to a scrubbed and organized Fixed Data table can be tortuous. It may take several steps simply to get the source data loaded into the Raw Data tables. The source data may not originally be in an electronic form. Even when it is, the format may not be one that can be directly imported into a spreadsheet.

In this section I will discuss the different ways that data can be entered into a spreadsheet, their pros and cons, specific issues and how to make them as efficient as possible.

PAPER SOURCE DOCUMENTS

If the source data is not already in an electronic form the user has two choices:

- Key the data into the Raw Data table
- Scan the source document

If there is a small volume of data to enter, keying might be the best approach; if there is a large volume it is probably desirable to scan it. Scanning isn't a "free" solution – it will be necessary to manipulate the scanned file before it can be imported into the spreadsheet.

Most scanning equipment and software can produce either an image file (such as a .jpg or .png) or a PDF. It is generally preferable to save the scan as a PDF if the user has access to Adobe Acrobat. The subscription versions of Acrobat have features to automatically convert a PDF file into various Microsoft Office file formats (Word, PowerPoint, RTF, Excel).

The degree to which the converted PDF file can be directly used will depend on the characteristics of the source document and the specific needs of the application. The source data spreadsheet resulting from the conversion may well require some adjustment to get it into a table format that can be imported into the Raw Data table of the main spreadsheet.

If a full-function version of Acrobat is not available it is possible to extract the data from the basic PDF reader and get it into a spreadsheet. Once the data is in a spreadsheet it can be adjusted into a usable form

To extract the data from the PDF do the following:

1. Open a blank spreadsheet file
2. Open the PDF file with the PDF reader
3. From within the PDF reader click on the Edit tool
4. Click on "Select All"
5. Click on "Copy
6. Go to the blank spreadsheet
7. Click on cell A1 (or the next empty row in column A)
8. Enter <ctrl>+V (paste)
9. Repeat steps 3 through 8 for each page of the PDF

The contents of the PDF will be pasted into the spreadsheet, row by row and page by page, into column A. The desired data in can be found and extracted ("parsed") from each row and stored into separate cells. As discussed previously, using a fixed-width font such as Courier New will make the parsing easier. When the required data has been parsed the resulting table can be imported (copied or moved) into the Raw Data table of the main spreadsheet.

This process is illustrated in the following screen shots. For simplicity, only the first page of the report is extracted.

FIGURE 10. EXTRACTING DATA FROM A PDF FILE

Typical business financial management report (step 2):

Fiscal Year to Date Financial Analysis

Financial Summary by Category & Subcategory - All Numbers in 000s

Classification		Previous Month				Current Month				Fiscal Year to Date			
Category	Subcategory	Actual	Disc	Budget	Percent	Actual	Disc	Budget	Percent	Actual	Disc	Budget	Percent
Data Through April	Grand Totals - Rev:	96,333	11,894	84,439	114.1%	50,348	(43,532)	93,880	53.6%	501,704	(81,856)	583,560	86.0%
	Exp:	(91,792)	752	(92,545)	99.2%	(48,439)	38,980	(87,419)	55.4%	(519,024)	99,546	(618,570)	83.9%
	Net:	4,540	12,646	(8,106)	-56.0%	1,909	(4,552)	6,461	29.5%	(17,320)	17,690	(35,010)	49.5%
	Revenues												
00a. Investments	1 Brokerage Accounts	672	172	500	134.4%	469	219	250	187.6%	4,102	1,422	2,680	153.1%
	2 Bank Accounts	14	(6)	20	70.0%	23	3	20	115.0%	133	(7)	140	95.0%
	3 Inventory Swaps	4	(6)	10	40.0%	12	2	10	120.0%	59	(11)	70	84.3%
00b. Direct Sales	1 East	28,296	(1,704)	30,000	94.3%	26,217	(8,783)	35,000	74.9%	184,768	(27,602)	212,370	87.0%
	2 West	18,691	3,691	15,000	124.6%	21,277	3,277	18,000	118.2%	133,854	26,124	107,730	124.2%
	3 South	44,518	8,918	35,600	125.1%	420	(36,580)	37,000	1.1%	157,813	(78,767)	236,580	66.7%
	4 International	1,247	(263)	1,510	82.6%	614	(1,486)	2,100	29.2%	6,674	(5,996)	12,670	52.7%
00c. Indirect Sales	1 Licensed Dealers	2,635	1,135	1,500	175.7%	897	(603)	1,500	59.8%	11,847	1,627	10,220	115.9%
	2 Third Party	256	256		N/A	419	419		N/A	2,454	2,454		N/A
	Expenses												
01a. Tax	1 Income - Federal	(2,890)	(366)	(2,524)	114.5%	(1,510)	1,306	(2,816)	53.6%	(15,848)	3,202	(19,050)	83.2%
	2 Sales - State	(3,660)	(436)	(3,224)	113.5%	(3,354)	2,946	(6,300)	53.2%	(23,677)	8,843	(32,520)	72.8%
	3 Property Tax	(1,205)	(5)	(1,200)	100.4%	(1,500)	300	(1,800)	83.3%	(9,438)	1,562	(11,000)	85.8%
01b. Financing	1 Construction in Process	(500)	(20)	(480)	104.2%	(516)	(16)	(500)	103.2%	(3,428)	202	(3,630)	94.4%
	2 Construction Completed												
01c. Utilities	1 Communications	(139)	11	(150)	92.7%	(162)	(12)	(150)	108.0%	(1,014)	46	(1,060)	95.7%
	2 Electric	(72)	57	(129)	55.6%	(83)	26	(108)	76.2%	(565)	205	(770)	73.4%
	3 Gas	(122)	(71)	(51)	239.5%	(44)	39	(83)	52.8%	(591)	(151)	(440)	134.3%
	4 Sewer	(21)	4	(25)	84.0%	(30)	3	(33)	90.9%	(177)	33	(210)	84.3%
	5 Water	(31)	(10)	(21)	148.9%	(44)	(15)	(29)	151.7%	(277)	(107)	(170)	162.9%
02. Insurance	1 Property	(44)	1	(45)	97.8%	(48)	3	(51)	94.1%	(329)	21	(350)	94.0%
	2 Employee Medical	(218)	(18)	(200)	109.0%	(226)	(13)	(213)	106.1%	(1,588)	(238)	(1,350)	117.6%
	3 Vehicles	(261)		(261)	100.0%	(273)		(273)	100.0%	(1,999)	1	(2,000)	100.0%
03. Automobile	1 Gasoline	(120)	30	(150)	80.3%	(102)	44	(140)	69.9%	(733)	337	(1,070)	68.5%

Page 1 of 1

Spreadsheet column A after the data has been copied from the PDF (steps 3-5) and pasted into the spreadsheet (steps 6-8):

	A
1	Fiscal Year to Date Financial Analysis
2	Category Subcategory Actual Disc Budget Percent Actual Disc Budget Percent Actual Disc Budget Percent
3	Data Through Grand Totals - Rev: 96,333 11,894 84,439 114.1% 50,348 (43,532) 93,880 53.6% 501,704 (81,856) 583,560 86.0%
4	April Exp: (91,792) 752 (92,545) 99.2% (48,439) 38,980 (87,419) 55.4% (519,024) 99,546 (618,570) 83.9%
5	Net: 4,540 12,646 (8,106) -56.0% 1,909 (4,552) 6,461 29.5% (17,320) 17,690 (35,010) 49.5%
6	Revenues
7	Financial Summary by Category & Subcategory - All Numbers in 000s
8	Classification Previous Month Current Month Fiscal Year to Date
9	00a. Investments 1 Brokerage Accounts 672 172 500 134.4% 469 219 250 187.6% 4,102 1,422 2,680 153.1%
10	2 Bank Accounts 14 (6) 20 70.0% 23 3 20 115.0% 133 (7) 140 95.0%
11	3 Inventory Swaps 4 (6) 10 40.0% 12 2 10 120.0% 59 (11) 70 84.3%
12	00b. Direct Sales 1 East 28,296 (1,704) 30,000 94.3% 26,217 (8,783) 35,000 74.9% 184,768 (27,602) 212,370 87.0%
13	2 West 18,691 3,691 15,000 124.6% 21,277 3,277 18,000 118.2% 133,854 26,124 107,730 124.2%
14	3 South 44,518 8,918 35,600 125.1% 420 (36,580) 37,000 1.1% 157,813 (78,767) 236,580 66.7%
15	4 International 1,247 (263) 1,510 82.6% 614 (1,486) 2,100 29.2% 6,674 (5,996) 12,670 52.7%
16	00c. Indirect Sales 1 Licensed Dealers 2,635 1,135 1,500 175.7% 897 (603) 1,500 59.8% 11,847 1,627 10,220 115.9%
17	2 Third Party 256 256 N/A 419 419 N/A 2,454 2,454 N/A
18	01a. Tax 1 Income - Federal (2,890) (366) (2,524) 114.5% (1,510) 1,306 (2,816) 53.6% (15,848) 3,202 (19,050) 83.2%
19	2 Sales - State (3,660) (436) (3,224) 113.5% (3,354) 2,946 (6,300) 53.2% (23,677) 8,843 (32,520) 72.8%
20	3 Property Tax (1,205) (5) (1,200) 100.4% (1,500) 300 (1,800) 83.3% (9,438) 1,562 (11,000) 85.8%
21	01b. Financing 1 Construction in Process (500) (20) (480) 104.2% (516) (16) (500) 103.2% (3,428) 202 (3,630) 94.4%
22	2 Construction Completed
23	01c. Utilities 1 Communications (139) 11 (150) 92.7% (162) (12) (150) 108.0% (1,014) 46 (1,060) 95.7%
24	2 Electric (72) 57 (129) 55.6% (83) 26 (108) 76.2% (565) 205 (770) 73.4%
25	3 Gas (122) (71) (51) 239.5% (44) 39 (83) 52.8% (591) (151) (440) 134.3%
26	4 Sewer (21) 4 (25) 84.0% (30) 3 (33) 90.9% (177) 33 (210) 84.3%

Formulas to parse the various fields added to columns C through L (column A shrunk for clarity):

	A	B	C	D	E	F	G	H	I	J	K	L
1	Fiscal Year to I		Detail?	Catg?	Category	Sub Category	Prev Act	Prev Bud	Curr Act	Curr Bud	YTD Act	YTD Bud
2	Category Subcate		0	0								
3	Data Through Gra		0	0								
4	April Exp: (91,7		0	0								
5	Net: 4,540 12,64		0	0								
6	Revenues		0	0								
7	Financial Summar		0	0								
8	Classification		0	0								
9	00a. Investments		1	1	00a. Investments	1 Brokerage Accounts	672	500	469	250	4,102	2,680
10	2 Bank Accounts		1	0		2 Bank Accounts	14	20	23	20	133	140
11	3 Inventory Swa		1	0		3 Inventory Swaps	4	10	12	10	59	70
12	00b. Direct Sale		1	1	00b. DirectSales	1 East	28,296	30,000	26,217	35,000	184,768	212,370
13	2 West 18,691 3,		1	0		2 West	18,691	15,000	21,277	18,000	133,854	107,730
14	3 South 44,518 8		1	0		3 South	44,518	35,600	420	37,000	157,813	236,580
15	4 International		1	0		4 International	1,247	1,510	614	2,100	6,674	12,670
16	00c. Indirect Sa		1	1	00c. Indirect Sales	1 Licensed Dealers	2,635	1,500	897	1,500	11,847	10,220
17	2 Third Party 2		1	0		2 Third Party	256		419		2,411	
18	01a. Tax 1 Incom		1	1	01a. Tax	1 Income - Federal	(2,890)	(2,524)	(1,510)	(2,816)	(15,848)	(19,050)
19	2 Sales - State		1	0		2 Sales - State	(3,660)	(3,224)	(3,354)	(6,300)	(23,677)	(32,520)
20	3 Property Tax (		1	0		3 Property Tax	(1,205)	(1,200)	(1,500)	(1,800)	(9,438)	(11,000)
21	01b. Financing 1		1	1	01.b Financing	1 Construction in Process	(500)	(480)	(516)	(500)	(3,428)	(3,630)
22	2 Construction (		1	0		2 Construction Completed	0	0	0	0	0	0
23	01c. Utilities 1		1	1	01c. Utilities	1 Communications	(139)	(150)	(162)	(150)	(1,014)	(1,060)
24	2 Electric (72)		1	0		2 Electric	(72)	(129)	(83)	(108)	(565)	(770)
25	3 Gas (122) (71)		1	0		3 Gas	(122)	(51)	(44)	(83)	(591)	(440)
26	4 Sewer (21) 4 (		1	0		4 Sewer	(21)	(25)	(30)	(33)	(177)	(177)

WEB SOURCE DOCUMENT WITHOUT EXPORT CAPABILITY

Many web sites provide an export capability to extract data and store it in a spreadsheet-compatible file format. However, sometimes the desired data isn't included in the export. Or, the web site may not provide this capability.

A common example of this is with bank account and credit card account web sites. These web sites will usually list both pending and posted transactions in two different sections on the web page. However, the export function only provides the posted transactions. Both sets of transactions may be needed to get a complete picture of the financial situation.

Usually the transaction sections of these web pages are built using some type of SQL table, so extracting the data from them is relatively easy. To do this, highlight (select) all the desired cells and copy them to the clipboard (<ctrl>+C). Go to a designated spot in the Raw Data table[33] and paste the values from the clipboard. Since the data in the clipboard isn't from another spreadsheet, Excel won't allow paste > special > values. Instead it will provide the option to do paste > special > text.

When the data is pasted into the spreadsheet each cell in the web site table will be pasted into a separate cell in the spreadsheet. Depending on how the web page was built the data may or may not be pasted into a table with the same format (rows and columns) as the web page. If it is, your

---

[33] In this example the Spreadsheet will have (at least) two Raw Data tables – one for the exported posted transactions, and one for the copied pending transactions.

work is done: Life is good; go out and celebrate. If it is not you will have to reformat it. Often the data will be loaded into successive cells in a single column. If the original table consisted of *M* rows, each containing *N* columns, after the paste the spreadsheet table would contain *M*N* rows by 1 column.

Since the structure of the table is consistent and each data element is in a separate cell it is easy to convert the pasted table back into the form of the original web page table. This can be accomplished by doing the following:

1. Pick a field in the web page table that is guaranteed to be distinguishable from all the other columns. For these kinds of financial applications I usually use the Transaction Date field. It will always be a numeric value and within a known range of values.
2. Use this field to set a "Valid?" flag for the rows where the cell satisfies the criteria defined in step 1. This flag will denote a defined and consistent position in every row of the original web page table.
3. For each "Valid" row, use appropriate formulas to extract data from the rows above or below the "Valid" row as necessary. If the row is not "Valid", set the cells in that row to blanks or zeros.

The result of this is shown in the following example of a download from a company credit card web site:

FIGURE 11. EXTRACTING DATA FROM A WEB SITE

B3 =IF(AND(ISNUMBER(A3),A3>B1-20,A3<=B1),1,0)

	A	B	C	D	E	F
1	Control Date:	04/12/20			Raw Data Table	
2	Pending Transactions Download	Valid?	Type	Date	Description	Amount
3	43,928.00	1	Sale	04/07/20	FRED'S STATIONARY MD FREDERICK	-58.82
4		0				
5	FRED'S STATIONARY MD FREDERICK	0				
6	58.82	0				
7		0				
8		0				
9	43,928.00	1	Sale	04/07/20	SAM'S CLUB 00841 GA ATLANTA	-69.30
10		0				
11	SAM'S CLUB 00841 GA ATLANTA	0				
12	69.30	0				
13		0				
14		0				
15	43,928.00	1	Sale	04/07/20	SHELL TX HUMBLE	-13.04
16		0				
17	SHELL TX HUMBLE	0				
18	13.04	0				
19		0				
20		0				
21	43,929.00	1	Sale	04/08/20	THE HOME DEPOT #2911 GA ATLANTA	-5.32
22		0				
23	THE HOME DEPOT #2911 GA ATLANTA	0				
24	5.32	0				
25		0				

The criteria to determine which row will be flagged as "Valid" can be seen in the formula bar. In this example, the Date field is being used as the control. The test used to identify the date field is that it must be numeric, greater than the value of the control date minus 20 (i.e. not more than 20 days prior) and less than or equal to the value of the control date.

It could be possible that there is an amount field on a transaction that falls into that range and would therefore incorrectly be identified as a date. This is something that must always be considered and protected against. In this example the developer knows that the dollar limit on these transactions is low enough that this will never happen. (The numeric value of current dates are greater than 44,500. The maximum allowed transaction value is $25,000.) I will say it again: Understanding the data is an essential requirement and the first step to be performed when developing a spreadsheet.

MERGING MULTIPLE OR CUMULATIVE DATA SOURCES

Not every spreadsheet application starts from scratch every time it is used. Some applications are intended to be cumulative. An analogous situation exists when the data for a spreadsheet is being imported from multiple sources. These situations create special challenges, and both are typically handled in the same way.

In either situation (or in the situation where the spreadsheet needs to support both) the source data load processes will involve three steps rather than the two steps that have been discussed up to now. The first step (importing the data into the Raw Data table) and the second step (scrubbing and organizing the data into the Fixed Data table) are the same. The third step involves populating a Master Data table where the data from multiple sources and / or multiple updates is accumulated. The Master Data table will typically consist of two areas (i.e. sets of columns): a "data area" and a "test area". Like the Raw Data table, the data area of the Master Data table contains values, not formulas. Unlike the Raw Data table, the Master Data table is not cleared before each update. The test area contains formulas similar to the Fixed Data table. These formulas can perform a variety of tests, bring in data from reference tables, etc. Two important tests that should always be included are to identify valid rows ("Valid?") and potential duplicates ("Dup?").

Important considerations that should be followed for this approach to work.

- If there are multiple data sources, the data from each source must be transferred into the data area of the Master Data table in the same format. Typically the data will be copied from the Fixed Data tables. For convenience, create an area in the Fixed Data tables that has the same structure (i.e. column definition) as the data area of the Master Data table.

- The data must be copied into the data area of the Master Data table as values. The data portion of the Master Data table can't contain formulas that reference the Fixed Data table(s) because the Fixed Data tables will change every time new data is imported.

- The Master Data table must be designed large enough (i.e. have enough rows) to accommodate all the data that will be imported from all runs and all data sources during the spreadsheet's expected lifetime. How many rows that requires will obviously depend on the needs of the specific application. Since an Excel table can now have over one million rows there is no need to be stingy[34].

- The Master Data table should include a combination of columns that can be used as a unique key so it can be sorted. This is so the rows can be tested for potential duplicates. After a new batch of input has been copied into the Master Data table, sort the table. If there is overlap in the data consecutive rows will be flagged as duplicates. Examine these rows. If they are duplicates clear the data portion of one of them[35]. If they aren't

---

[34] In some of my applications I allocate 50,000 rows per data source.

duplicates determine why they have the same key. This might indicate that the chosen key structure isn't sufficiently unique and needs to be enhanced.

---

[35] Clear the data portion, not the formulas. If you clear the formulas they won't be available for whatever data gets loaded into that row on the next run. Also, do not delete the row. If you do Excel will have to update all of the affected cell references, which can take a while on very large tables. It will also create #Ref errors on any formulas that reference that row (such as the duplicate check).

7. TRANSFORMATION: THE STRUCTURE OF MAGIC

In the last chapter we examined how to recognize, evaluate and import the data content of the problem. If data is the heart of a spreadsheet then the processes whereby the data is transformed into meaningful information is its soul.

Sometimes the transformation process is simple and straight forward – select the transactions of interest and add them up to get a total. This type of simple transformation can be found in almost every spreadsheet. SUM is probably the single most used of all spreadsheet functions. It was one of the first.

Other transformations may be extremely complex and subtle – optimizing a process or schedule for example. Sometimes these transformations can be so esoteric that they do indeed appear to involve magic!

Regardless of the type of transformation being performed, there are several basic guidelines that should be followed:

- Just as the characteristics of the original data must be known and defined, the characteristics of the final information must also be known and defined. Developing an explicit, unambiguous, and actionable definition of the desired end result is often the most difficult challenge faced by the developer of a spreadsheet.

- The steps of the transformation process must be defined. What is the relationship between the transaction data and the end result(s)? In what way must the transaction data relate to the reference data (if any)? Given that each table can (or should) only perform a single step in the transformation process, how many interim transformation tables will be needed?

- What "operational parameters" must be specified? What types of selection criteria must be specified? Will all data be loaded each time the spreadsheet is updated, or is the data cumulative (retained and added to each time)?

All these questions must be answered.

If this sounds a lot like the activities that developers go through when creating a program using a traditional programming language, it is. These are all elements of the problem that must be defined in order to create a workable solution. There is no way around it. The good news is that spreadsheet software provides capabilities which make this process significantly easier. A partial or incomplete definition can be tried and then easily refined until the desired results are obtained. Common errors (such as neglecting to freeze a cell reference when copying a formula) are usually glaringly obvious. Data representing boundary conditions[36] can be entered and tested to ensure they are handled correctly.

Perhaps the most important attribute of spreadsheet software is that it provides immediate feedback. The

---

[36] Huh?? What's a boundary condition? We'll get to that in a bit. Or a byte. Or whenever.

developer actively interacts with the computer in real time. This makes it feasible for the user and the developer (if they are not the same person) to sit together and jointly prepare the solution. This *collaborative development* has been shown to be an extremely effective approach to problem solving.[37]

[37] And there a tools available to facilitate this collaboration. These, however, are outside of the scope of this discussion.

Roots

Before a definition of the solution can be developed there must be agreement on the terminology. This is particularly important where the developer of a program is not the same person who will ultimately use the output of that program, regardless of the language it is written in. Differences in interpretation and understanding have haunted the field of programming since its earliest days and are the root-cause of many so-called bugs[38]. How many organizations base important decisions on reports where the exact meaning of the terms used in the reports are different than the understanding of the managers using them?

Lewis Carol may have said it best:

> *"When I use a word, Humpty Dumpty said in a rather scornful tone, it means just what I choose it to mean – neither more nor less."*

Traditional programming efforts have attempted to circumvent this problem by requiring the user (or more often a delegate) to prepare a detailed specification. The specification document, which can run into hundreds of pages, is provided to the programming staff. The programmers working on the assignment frequently have little knowledge of the user's business but do their best to

---

[38] The use of the term "bug" to refer to a problem in a computer program originated on the 9th of September 1947 when a moth got stuck in a relay in an early computer (the Mark II Aiken Relay Calculator at Harvard) causing the program to fail. So the first "computer bug" really was an insect! FYI – the moth was taped onto the computer log and can still be viewed in the archives of the Smithsonian (and also on page 306).

interpret what they have been provided. What typically follows are numerous rounds of testing, error reports, and program "fixes". The usual end result of this exercise is that the programming project takes longer than expected and the "differences in interpretation" are sufficiently well hidden that they are not obvious to either the testers or the ultimate user.

Not a good approach.

Some companies have attempted to develop "Data Dictionaries" to explicitly define exactly what each data element and term used in their reports mean. This effort has usually met with limited success. Developing an organization-wide data dictionary can be a daunting (read: expensive) task with little quantifiable ROI. If it is not actively maintained the dictionary quickly becomes obsolete and therefore less than useful and usually ignored.

Good luck.

The biggest challenge with developing a set of common definitions is that, for very legitimate business reasons, different groups want to see things differently to effectively manage their part of the business. I had direct experience with this at a large chain drug store company where I once worked.

This retail drug store's business was managed as two major divisions: OTC (over the counter) and Rx (prescriptions). The problem was that the marketing and sales organizations viewed the businesses quite differently. To the marketing organization, OTC & Rx referred to how the product being sold was purchased. To the sales organization, it referred to how the sale was recorded. If the item was sold at the cash register it was considered OTC; if the item was put in a prescription bag it was Rx. This difference led to endless wasteful discussions as to how each of the businesses was really doing.

To resolve the disputes, I developed a summary P&L report that had four quadrants on one page. Vertically, the two sections showed the Sales, Cost, Gross Profits and other key values for items purchased as OTC and Rx. Horizontally, the two sections showed the same data for items sold at the cash register (OTC) or at the pharmacy (Rx), There were grand totals for rows and columns. Not surprisingly, the grand totals for both were the same. Of course, the report showed that some areas were performing better than they had claimed and some worse. Depending on which, the affected executives either liked the report or hated it (and me). Such is the world of corporate politics.

Little Steps, Big Steps

The transformation steps of a spreadsheet define the *relationship* between the final information and the original data. Each step must be represented as a formula in a cell. Defining the formulas for each step in the transformation process is the central challenge when developing a complex spreadsheet.

As discussed in Chapter 3, each cell in a spreadsheet has two personalities: its contents and its value. Each step of the transformation process consists of cells yielding values derived from the formulas specified in their contents. Each of these formulas consists of one or more functions whose parameters are often the values of other cells. This process can repeat multiple times in a cascading fashion as is shown in this illustration:

FIGURE 12. CASCADING CELL REFERENCES

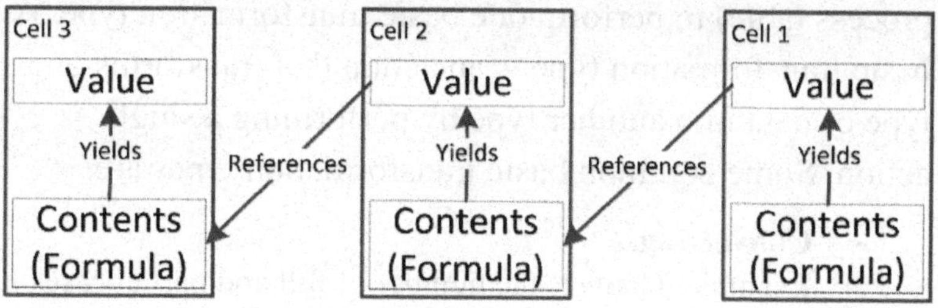

Formulas are made of combinations of functions. Functions take values for each of their parameters and yield a value as the result. Therefore, any function that yields a value of the correct data type can be used as a parameter to another function. As a result, there is no practical limit on how complex a formula can be.[39]

A fundamental question to address when designing a spreadsheet is to decide how many steps are needed to completely specify the transformation. The general rule is the fewer the steps, the more complex each formula will need to be. A key consideration when making this decision is maintainability. Extremely complex formulas can be difficult to verify (debug) and maintain (update, implement new business rules), especially if the person doing the maintenance is not the original developer. Sometimes the position of the cell is important because it is part of the final report. In this case, the formula may have to be complex to produce the correct value. This is the case in the project Gantt chart that is discussed in a later chapter. If the position of the cell is not critical then it is probably better to break the transformation process into multiple cascading steps as is shown in the previous diagram.

An extreme approach is for each step of the transformation process (i.e. the formula in each cell of a transformation process table) to perform one basic transformation type. A basic transformation type is an action that transforms one type of data into another type by performing a single action. Some common basic transformation types are:

- **Converting**.
  Example: Convert the number of full and partial cases in the Order Data table to units by multiplying the Order Quantity by the Units per Case.

- **Associating**.
  Example: Lookup the Customer ID in the Order Data in

[39] There is a limit (any individual cell can contain up to 32,767 characters), but your brain will probably explode before you get there.

the Customer Reference Table to get the State the customer is located in

- **Selecting.**
  Example: If the order quantity is positive then make the Order Type "Sale", otherwise "Return".

- **Filtering**.
  Example: Select only the customers located in California.

- **Consolidating**.
  Example: Sum the Units Ordered for the selected customers to get a total for the state.

This is not an exhaustive list, but you get the idea.

Doing this may make the spreadsheet larger, but it will simplify the debugging and maintenance later on. What basic transformation types are needed, and in what order they need to be performed, are entirely dependent on the problem being solved. In practice the spreadsheet developer must find an appropriate balance between many simple formulas and a few complex ones.

Once the decision is made as to how many different steps the transformation process will require, and therefore how many different formulas need to be written, the next question is where to put them.

There are two general ways the formulas can be laid out:

- If the data in the table the formulas reference is one dimensional (i.e. each unique data element is in a single column), then it may be best to place the formulas side by side. For example: The formula in cell A6 uses the value result of the formula in cell B6; the formula in cell B6 uses the value result of the formula in cell C6; etc. Since most tables are laid out vertically (i.e. each row represents a record), the formulas that act on a particular data element in each of those rows should also be laid out vertically. Once the formulas are written they can be copied down to reference every row in the table.

- If the data in the table the formulas reference is two dimensional (i.e. the same type of data element is in multiple columns) then the formulas might be placed in separate tables on separate tabs. This is frequently seen in applications where the rows of the table represent one attribute of a problem (such as location) and the columns represent another (such as time). Every cell in the table would have the same type of data (such as temperature at that location at that time). The tables in each tab would all have the same basic layout (row and column definition) and the cells would contain the formula for one step of the transformation. Stacking the transformation formulas in separate tabs like a layer cake makes the process much easier to understand and debug.

- Sometimes the problem being solved involves more than two dimensions. For example, the models used to forecast the weather break the atmosphere into discreet nodes (the current climate models break the earth into square "blocks" 100 kilometers on a side) and specify the location (latitude and longitude) and vertical height of the node. For each node the model considers the temperature, humidity, wind direction and speed, pressure, etc. at that node at a specific time. Complex formulas take this information and project it forward in time.

  Problems like this are not good candidates for being solved using a spreadsheet – I would advise against

using a spreadsheet to predict the weather!
This is however a very important point: Understanding the *dimensionality* of the data is one of the key considerations when deciding how to address a problem, and which tool to use.

Imposing Structures

In its most basic sense, a spreadsheet can be viewed as a collection of tables, and formulas that perform operations on the data in those tables. There are many similarities between a spreadsheet and a traditional procedural program. A procedural program has executable code that performs operations on the data in files, or more often now, database tables. Unlike procedural languages, where the lines of code operate on all the appropriate cells in the table, spreadsheet formulas are contained within each individual cell of the tables themselves. As with any program, certain practices should be followed in the design of the tables and formulas in a spreadsheet to ensure that it is efficient, yields the correct results, and is maintainable.

TABLES

There are many analogies between a relational database table and a spreadsheet table. When creating a relational database table, the following general principles apply:

- Each column of the table represents a unique data element

- Each column (data element) must have a unique name to identify it.

- Each row of the table represents a unique occurrence of the collection of data elements that are considered as a group or unit. This is usually referred to as a record.

- The size of the table is determined by the number of columns and the number of rows it contains. More columns and / or rows, bigger table.

Unlike a relational database table though, these are not hard constraints when creating a spreadsheet table:

- Multiple columns in the table can represent the same data element.

- The columns are not required to have a unique name, or any name at all (though it is recommended that they do).

- The size of the table is fixed and defined as part of the spreadsheet development when the table is created.[40]

As with so many things in life, just because you can doesn't mean you should. Like a relational database table, a table in a spreadsheet should usually follow the same general principles unless there is a legitimate reason not to do so. There are times when it is desirable to violate some of the

---

[40] There are ways around this constraint, but they often impose other restrictions that make them less than optimal. For example, a named range can be defined using a formula with the OFFSET function, but this requires that all the data in the table be contiguous. The Excel "Table" feature (Insert > Table) can be used to create a dynamic table, but these tables also have limitations and restrictions. None of these solutions circumvent the need to allocate enough rows to handle the maximum amount of data that may be imported (and test for possible overflow). Perhaps the most important consideration is that using esoteric capabilities like these can create obscure errors and make the spreadsheet significantly harder to debug and maintain. Always remember the first law of programming: **KISS** (Keep It Straight-forward, Stupid).

general design guidelines. Sometimes "bending the rules" can allow the spreadsheet developer to accomplish something that would not otherwise be possible, or in a more efficient and straight-forward way, What liberties the developer may want to take will depend on the purpose of the table (source data, transformation, presentation), the type of data in the table (transaction, reference, other), and the specific needs of the problem being addressed. What is important is that it is done deliberately, not because the developer is lazy or sloppy.

Headers: One of the principles that should always be followed is to put a *meaningful header* at the top of every table. The only time this might not be possible is when source data must be moved into a Raw Data table. In this situation the developer of the spreadsheet may have no control over the structure or content of the data, and just has to use what is provided.

A meaningful header should have most or all of the following components:

- It should identify every column in the table with a brief (single cell) but meaningful name.
- The column names should be unique and at the top of the table, usually on Row 1.
- All the header rows should be included in the "named range" for the table.

- The row directly below the column names should have an offset counter for each column of the table
- It should have one or more rows that can be used to display meaningful totals (SUM, COUNT, etc.) of various fields in the table.
- It should have a blank row as the last row of the header.
- All of the rows in the header should be "frozen" (View > Freeze Panes) so that they can be seen and accessed as the user scrolls through the data in the table.

This may seem a bit excessive, but there are good reasons why each of these is important:

- Obviously, it is important to have a meaningful name for each column (data element) in the table so that the user and subsequent maintainers of the spreadsheet know what they represent.
- By putting the header at the top of the table and making each name unique the table can be used as the source data for a PivotTable. Including the header as part of the Named Range associated with the table is required for the same reason (assuming you follow the good practice of using named ranges for PivotTables). The other rows in the header can be excluded from any PivotTable by flagging them as not valid (the cell in the "Valid?" column for those rows can be set to 0 or FALSE) and using the "Valid?" field as a PivotTable Filter criteria.
- The offset counters in the row directly below the column names show the relative number of each column in the table, starting with the leftmost column of the table as 1. These counters facilitate accessing the fields in the table

using the lookup functions.
The lookup functions (VLOOKUP, HLOOKUP) provide a spreadsheet with capabilities similar to the JOIN operation in a relational database. With a lookup function, a formula in a cell in one part of a spreadsheet can dynamically access (meaning it will change depending on what data has been imported) cells in a table in another part of the spreadsheet. The formula simply has to specify the key value of the record it is looking for, what table to search, and where in the table the desired field is located.[41]
It is this last part that can create problems. A spreadsheet can't identify a specific field in a table by name as in a relational database[42]. Identifying where the desired field is located in the table is done by specifying the offset of where the field is located. This is the relative column number (if using VLOOKUP) or relative row number (if using HLOOKUP) of the field. Problems arise if the table is modified by subsequent maintenance activities. If columns are added or deleted the offset of the desired field will change. If the offset is hard coded in the formulas that access the table the VLOOKUP will return the wrong value. The same problem will occur for an HLOOKUP if rows in the table being searched are added or deleted.
Fortunately there is a simple way to avoid this problem. Instead of the formulas specifying a hard coded offset, The offset can be determined dynamically each time the spreadsheet is updated using the name of the desired field and a pair of lookup functions.

---

[41] The equivalent SQL statement would be: SELECT <desired field name> FROM <table name> WHERE <key field name> = <specified key value>

[42] The Excel Table feature supports referencing fields by name, but as mentioned earlier, this has limitations and restrictions that limits its usefulness.

To illustrate how this works, consider the following example: A spreadsheet has an order table (named Ord_Tbl) and a customer table (named Cust_Tbl). Each record in the order table has a customer number field (which happens to be in column "D" of the order table), and the developer wants to populate the customer name in column "R" of the order table. The name associated with the customer name field in the customer table (i.e. in the first row of the header) is "Cust_Nm". This field is in column "H", which is the 8th column from the start of the table. The formula to get the customer name (in cell "R16" of the order table in this example) would look like this:

```
=VLOOKUP(D16,Cust_Tbl,8,0)
```

If changes are made to the Customer Table, the column where the customer name is located may change, and the formula will no longer work. Instead of coding an offset of "8" in the lookup function, the developer codes the following formula:

```
=VLOOKUP(D16,Cust_Tbl,HLOOKUP("Cust_Nm",Cust_Tbl,2,0),0)
```

The HLOOKUP function retrieves the value in the second row of the customer table, which has the offset value for each field (column) in the table. Doing this enables the offset value to be retrieved using the name of the field. If the customer table is modified the returned offset value will change to reflect the new structure of the table (as long as the person doing the maintenance follows expected development practices and also updates the offset row[43]).

---

[43] There have been reports that maintenance developers who didn't follow these expected practices were dragged outside and flogged.

One final efficiency consideration: instead of coding the HLOOKUP in the formula on every row, the name of the desired field and the result of the HLOOKUP can be placed in cells somewhere else (perhaps in rows in the header), and all the rows refer to the value there. This saves doing numerous extra HLOOKUPs, and one place to change if it becomes necessary (such as a different field is needed).
Yup, magic!

- It is often useful to have totals of various fields in a table. This is especially true in the Fixed Data table(s) where the imported data is being validated and cleaned. These totals can provide the user with valuable information such as:
  - How many records have been loaded?
  - How many are valid?
  - How many potential duplicates?
  - How many errors in each of the fields being tested?

  People (and accountants) traditionally think of totals being at the bottom of a list or table. In a spreadsheet, it is often useful to put them at the top of the table in the header where they can be frozen and always visible as the user scrolls through the data in the table. This can be helpful as errors in the source data are corrected, since the error counts will update as each error is corrected.

- The Auto Filter feature (Data > Filter) is an extremely powerful tool for selecting the rows in a table based on some specified criteria. When filtering is activated on a table, filter drop-down buttons are placed on the right side of each cell in the top row of the specified range. This can be annoying, as the drop-down buttons will obscure part or all of each cell's contents.
  To circumvent this problem, leave a blank row as the

> last row of the header, and use this as the top row of the filter range specification. Since it is an empty row, there won't be anything for the filter buttons to obscure. Since it is part of the frozen header, the filter buttons will be visible and accessible regardless of where the table is scrolled.
> More magic!

Creating a meaningful header is often the most critical part of creating an effective table structure.

Mirroring Fields: As discussed above, in general each unique data element should be in a single column of a table. Sometimes bending this rule can make it feasible to accomplish some actions that would not be possible, or significantly more difficult, otherwise.

Consider the following example: The order table discussed above (Ord_Tbl) has columns for the order number (Ord_Ord), customer number (Ord_Cst) and product number (Ord_Prd) in that order. Using VLOOKUP it is possible to find the customer and product for any order given the order number. What if there is a need to find the order number for a specific customer or product?

One simple way to do this is to define additional named ranges (sub-tables) within the order table where the first column of the new table is the customer number column or the product number column of the order table, and the last column is the same as the last column of the order table. The new tables can be used with the VLOOKUP function to find data in the order table by customer or product.

The VLOOKUP function won't be able to access the order number field (Ord_Ord) using these tables because it is to the left of the start of the table. The Excel lookup functions don't allow zero or negative offsets[44]. This issue can be overcome by adding a column to the tables to mirror the order number. The cells in this column (Ord_OrdCpy) simply have a formula to copy the corresponding field in the order number column. The VLOOKUP functions can reference this field using the new sub-tables and the appropriate offset.

The appropriate offset can be found using the same technique described above by taking the difference between the offset value for the mirror field and the offset value for the field which is the start of the new table. The formula to find the offset used for getting the order number using the new customer lookup table would be:

```
=HLOOKUP("Ord_OrdCpy",Ord_Tbl,2,0)
-HLOOKUP("Ord_Cst",Ord_Tbl,2,0)+1
```

---

[44] Recently (the 2020 release) Excel introduced XLOOKUP, which is in essence a fully generic lookup function. XLOOKUP allows the developer to explicitly specify the column (or row) to search, the column (or row) to return, what to do if the desired entry isn't found and how to perform the search. While providing a lot of capability and removing some of the limitations of VLOOKUP and HLOOKUP, it still must be used with care. As with the other lookup functions, changes to the size or layout (and especially the meaning of each column!) of a table may require changes to any XLOOKUP referencing it. Another consideration is that, because it is so flexible, the intent of the developer may not be intuitively obvious. This could create challenges during debugging or subsequent maintenance.
Another important consideration is compatibility. Not every user may have the latest version of Excel; not every user may be using Excel – there are other spreadsheet programs out there. Be careful when using the newest cool features!

Table Change & Overflow Testing: A complex spreadsheet may have dozens of interrelated tables. For some of these tables, it is critical that they have the same dimension as other tables. For example, the Fixed Data table(s) should always have the same number or rows as their corresponding Raw Data table(s). If they have fewer rows they may miss part of the imported data. If they have too many rows they may pick up irrelevant data that happened to be in rows below the end of the Raw Data table.

Periodic maintenance activities may be performed that change the size of the tables in a spreadsheet. This can occur, for example, if the Raw Data table needs to be expanded to accommodate a larger input file. Other tables may need to be expanded to accommodate the additional data. When this kind of maintenance is done, it is critical that all the interrelated tables be adjusted as necessary. If not, errors may result that can be extremely difficult to track down.

For this reason, it is a good practice for complex spreadsheets to include a "control area" where the key design indicators such as table dimensions are measured and compared to ensure that they are as expected. There are two simple ways to do this:

- The dimensions of a table that has a named range can be found using the Excel ROWS and COLUMNS functions. These functions take a range as the argument and return the number of rows or columns in the range respectively. These values can be compared to control values, or to the values returned by testing other tables, and if they are different display an error.

- As mentioned earlier, it is a good practice to put an indicator ("End of Table <whatever>") and a row of characters (I just use "x") across the bottom of a table. This serves two functions. If the table overflows, such as when data is copied or moved into a Raw Data table, it provides an obvious visual indicator that there is an issue. In addition, just as with the ROWS and COLUMNS test, there can be a test to ensure that the end-of-table indicator has not been overwritten.

Since these two tests actually serve separate and distinct functions, I recommend that the spreadsheet include both.

One final comment: Tables are created as part of the spreadsheet development process: A tab is selected. Each column is assigned a specific meaning. A maximum number of rows are chosen. A named range is associated with the table. Once the spreadsheet has been developed various user activities populate the tables with data.

The most obvious time a table is populated is when data is imported into the Raw Data tables (by whatever process). As discussed in the last chapter, the developer must know in advance what the maximum number of rows of data will be loaded to allow enough space. He / she won't know how many rows will be loaded each time the spreadsheet is updated, but that isn't important. Tables must be designed to handle the maximum expected volume of data.

The other time a table is populated is when a PivotTable is refreshed. A PivotTable can be refreshed when the spreadsheet is opened (by selecting the “Refresh data when opening file” PivotTable option), or manually (with the “Refresh All” PivotTable command). As with importing data, the space provided for the PivotTable must be large enough to accommodate the maximum expected volume of data, and controls must be built in to protect against overflows.

FORMULAS

As Dorothy discovered, there is often more than one road that will get you from the problem to the solution. Even after the path is clearly defined, there are still many decisions that must be made. Excel may provide many ways to accomplish any given objective. The challenge is picking the best one.

Often the challenge is to define what “best” actually means. If the users of the spreadsheet are relatively unsophisticated, best might mean making it as simple and fool-proof as possible. If the spreadsheet might need to be maintained frequently, best might mean carefully following good coding practices and extensive comments. If the problem being addressed is extremely large and complex, operational efficiency might be a key consideration. Depending on the characteristics of the problem and how the solution will be presented, maybe go with complex formulas that have multiple imbedded functions.

Whatever choices are made, the developer should never forget two important rules:

- A spreadsheet should yield useful information for any collection of valid data that is loaded into it.

- A spreadsheet should not require maintenance when new data is loaded unless the new data is *materially different* than the data that the spreadsheet was originally intended to accommodate. The material difference most often encountered (and most forgivable) is that the new data set is larger than what had previously been considered the maximum.

The awesome power of a spreadsheet lies in its ability to easily apply transformations (i.e. formulas) to a vast collection of data. A formula can be entered to use the data in a particular row of a table. With a simple copy/paste command, the same formula can be applied to the data in thousands of rows. Depending on the formula, specific cell references can be frozen so they always refer to the same cell. A good example of this is the reference to the lookup offset value in a header that was discussed earlier.

Which leads us to our third rule:

- Formulas should be *portable.* The mark of a good formula is that it can be copied to every row in a table and work correctly.

Boundary Conditions: At a deeper level, there is a fourth rule that that should always be followed:

- Formulas (individually or collectively) must work for every possible value provided.

Whoa! That sounds pretty intense.

Actually it's not as bad as it sounds. In ten words or less, what this means is that each formula (or group of formulas acting together) must be able to distinguish between legal illegal entries, and if legal, also correctly deal with what are often referred to as *boundary conditions*. Simply defined, a boundary condition in the context of a spreadsheet is the extreme legal value of any parameter to a function.

Still sounds pretty intimidating.

Actually its not. Testing for a boundary condition doesn't mean that every formula or function has to identify it as such. Depending on the formula, it might be OK for it to return an error condition such as #VALUE!, #N/A, etc. These indicate (glaringly!) that some bad data has been encountered. This is where it is useful to have control totals for certain fields at the top of a table. If the control total suddenly shows an error value then there is a problem somewhere in the data. The offending row(s) can easily be located by filtering on the error value in that column. The challenge is to identify when legal values are provided that yield illegal results.

If the values supplied are legal, why would the results be illegal? This is not a trivial question and is precisely the question that the developer must ask as he or she is creating the spreadsheet.

To illustrate this, consider a simple example: A common problem in project management is determining if a task is ahead or behind schedule by comparing the actual percent complete to the expected percent complete. The data for determining the expected percent complete usually comes from the project plan in the project management system. It contains several pieces of information including the task Id and description, planned start and completion dates, resource assigned, etc. The calculation for expected percent complete on a given task is quite simple:

= (Today's Date – Start Date) / (Finish Date – Start Date)

A spreadsheet is developed that imports the project plan data from the project management system, calculates the expected percent complete for each task, compares that value to the actual percent complete (possibly provided by the project billing system), and produces a report of tasks that are behind schedule. What could be simpler (or go wrong)?

Consider the following (and yes, I have encountered all of these):

- Is all the imported data valid? Project management systems often do a lousy job ensuring that the project plans are clean and accurate. When the data is imported it might include a lot of trash. A key job of the Fixed Data tables is to test the fields for validity (are required fields populated and contain the correct data type?).

- Are the supplied dates valid? Since dates are stored as numbers, any numeric value can represent a date. This doesn't necessarily mean that the dates provided are legitimate. Each task date should be compared against the planned start and finish dates for the entire project. Any task date that falls outside that range should be considered suspect.

- Is the finish date before the start date? This can (and does) happen, and will result in a negative expected percent complete.

These situations all fall under the general heading of "bad data." Any task that meets any of these conditions must be identified so the data can be corrected. There are other situations where the data is completely valid, but the results will still be invalid:

- Is the start date in the future? If the task isn't scheduled to start yet, the expected percent complete should be set to 0%, not calculated.

- Is the finish date in the past? If the task was already supposed to be complete, calculating the expected percent complete will yield a meaningless value greater than 100%. As above, in this situation the expected percent complete should be set to 100%.

- Are the start and finish dates the same? A short task might be scheduled to start and complete on the same day. If these dates are used in the formula it will produce a #DIV0! error. While it is easy to test for this situation, deciding what to return as the expected percent complete value is not so obvious. If the start, finish and today's date are all the same the formula should simply return some agreed-upon value, like 50%.

Another type of boundary condition can be seen when performing an operation on a series of data points. A common technique used in forecasting is to calculate a moving average. The moving average highlights trends in the data by replacing the data point for a specific day with the average of a certain number of days (for example, 7) either before, after or around the day under consideration.

This is an easy calculation that can be done using the AVG function. But what about the data at the beginning or end of the collection period? You can't calculate a backward-looking 7-day moving average for the first 6 days of the data, or a forward-looking average for the latest 6 days, Attempts to do so will at yield very misleading results.

These examples are not intended to be comprehensive. Its purpose is to demonstrate the kinds of questions that the developer must consider when building a robust spreadsheet.

Common Problems: The good news is that a spreadsheet, unlike a conventional program written in a traditional procedural programming language, won't abort when it encounters an error condition. The bad news is that a spreadsheet won't abort, so it isn't always obvious that an error has occurred.

Errors come in many flavors. Some errors occur as the spreadsheet is being developed or maintained and must be corrected immediately. If the developer supplies the wrong arguments to a function Excel will complain and won't accept the formula until it is corrected. Most errors, and the most subtle errors, are discovered as the spreadsheet is being used.

Some of these *execution errors* are obvious. A cell with a SUM formula showing an error value such as #Value! or #N/A is clearly an issue. Some aren't so obvious. The SUM is an actual number, but is it the right number?

Part of the art of developing a spreadsheet is recognizing where potential errors can occur and including <u>ways</u> to detect or prevent them.

Consider the following common situation: A report shows data for a variety of categories and subcategories. There is a subtotal for each category, and a grand total at the end. What is the best way to ensure that the subtotals and totals reflect the values in every row, but don't count any row more than once?

There are several ways this can be done, and some are more likely to result in errors than others. A PivotTable will do totaling and subtotaling automatically, but for a variety of reasons this option may not be desirable (raw PivotTables are usually pretty ugly).

If the totaling must be done manually, The most basic way, and the way most prone to errors, is to create a subtotal for each category, and have a formula in the grand total line the adds all of the subtotals. This is problematic for a couple of reasons.

- As subcategories are added, the totals and subtotals will have to be adjusted to include the new subcategories. If a new subcategory falls before the first row or after the last row of a SUM the SUM range might not pick it up.

- If categories are added or removed, the grand total will definitely have to be manually adjusted to include or exclude the new subcategory rows.

Probably the safest way is to use a PivotTable to create the list of categories, subcategories, subtotals and grand totals. Formulas can be placed next to the PivotTable fields that use the rows of the PivotTable as parameters in SUMIF functions to collect the appropriate values from the source tables.

Regardless of how the category and subcategory totals are developed, how can the user be assured that they are correct? Just as with the control totals used to ensure that the tables in a spreadsheet have been sized correctly, control totals can be used to ensure that the formulas in a report are handling the data correctly.

In this example, a control total can be calculated as a simple SUM of the relevant fields in all of the valid rows of the source tables regardless of category or subcategory. These control totals should equal the grand totals in the report. If every valid row of the source table should be included in a subcategory and category total, adding all of the rows in the report should be exactly three times the grand total amount (since each valid row of the source table will be included in a subcategory total, a category total, and the grand total).

Again, the purpose here is to demonstrate the kinds of questions that the developer must consider when building a robust spreadsheet.

Comments: Those of us who began life as programmers (sorry, developers) remember how important it is to document our programs. Typically, at least half the lines in a program were comments. This was to help the poor sap who had to maintain the program understand what the heck your twisted mind was thinking when you originally wrote it.

It is still a good idea to include comments in a spreadsheet. At one level, this is easy to do: Just type a description of a table or formula in any cell that isn't being used and that won't mess up your formatting. In a sense, that is exactly what you are doing when you create the header rows at the top of a table. You can also add a note to an individual cell to describe its purpose, function or whatever. In the collaborative versions of Excel you can add comments which can be replied to or added to by others.

But how do you document a formula, especially a really complex one? Notes and comments may be hard to relate to individual functions in the formula. There is also a significant risk that they won't get updated as the formula changes. What is needed is a way to include comments inside a formula itself.

Excel doesn't explicitly provide this capability, but there is a workaround. This takes advantage of the fact that a *mathematical identity* applied to a formula does not change the value of the formula. In other words, if you multiply a formula by one or add zero to it, its value doesn't change.

The Excel function "N" returns a value of zero if the parameter supplied to it isn't numeric. A comment can be supplied as the parameter to an N function and added to portion of a formula and it won't change the value returned by the formula.

Consider the resource scheduling formula from the earlier discussion about efficiency:

```
=
IF(ISTEXT($M120),0,
IF(Z$2<$M120,0,
IF(ISNUMBER($Q120),
ROUND(MAX(0,
MIN(
IF(ISNUMBER($P120),$P120,99999),
IF(ISNUMBER($O120),$O120*Z$3,99999),
IF(ISNUMBER($N120),$N120*VLOOKUP($Q120,Avail_Hrs,Z$1+3),99999),
IF(ISNUMBER($S120),$S120-SUM($V120:Y120),99999),
BG120,
IF(ISNUMBER($L120),VLOOKUP($L120,DepTbl,Z$1+1),99999)
)),0),0)))
```

Even with the <new line> characters inserted it is still unclear exactly what each part of the formula does. With N functions each part of the formula can be explained:

FIGURE 13. ADDING COMMENTS TO A FORMULA

```
=
IF(ISTEXT($M120),0+N("Task is on hold"),
IF(Z$2<$M120,0+N("Schedule date is before early start date"),
IF(ISNUMBER($Q120)+N("Resource has been assigned to this task"),
ROUND(MAX(0,
MIN(
IF(ISNUMBER($P120),$P120,99999)+N("Maximum allowed task hours"),
IF(ISNUMBER($O120),$O120*Z$3,99999)+N("Maximum allowed FTE"),
IF(ISNUMBER($N120),$N120*VLOOKUP($Q120,Avail_Hrs,Z$1+3),99999)+N("Maximum allowed percent"),
IF(ISNUMBER($S120),$S120-SUM($V120:Y120),99999)+N("Task hours remaining"),
BG120+N("Resource available hours"),
IF(ISNUMBER($L120),VLOOKUP($L120,DepTbl,Z$1+1),99999)+N("Task dependency constraint")
)),0),0)))
```

Still complex, but at least now the logic can be followed and evaluated.

ARRAY FORMULAS & OTHER CRIMES AGAINST NATURE

Like nature, the true beauty of spreadsheets lies in their ability to produce amazingly complex results from elegantly simple constructs. In nature, a few simple rules define how ice crystals must form as water molecules freeze. But the result is that no two snowflakes look exactly the same[45]. In a spreadsheet, simple formulas can be copied over many rows or columns to produce amazingly complex results.

Over the years since it was first introduced many new enhancements and features have been added to spreadsheet software. Most of these have been extremely helpful and have filled in gaps in the computational capabilities of the tool. Some, however, have apparently been added at the prodding of a small group of geeks whose sole mission in life seems to be demonstrating how clever they are. To my mind, they are a solution in search of a problem.

There are several reasons why I avoid Array Formulas and Excel's other more esoteric capabilities:

- For the average spreadsheet user they are hard to understand. This is something of a Catch-22: They are not as intuitive as most of the other spreadsheet functions. Therefore most spreadsheet developers won't take the time to really learn how to use them. Therefore they don't get used in most spreadsheets. Therefore they are encountered infrequently. Therefore there is seldom a need to really learn how to use them. Therefore most spreadsheet developers won't take the time to really

[45] Yes, I know this isn't strictly true, but it's what we were taught as children.

learn how to use them. And so it goes.

- As a direct consequence, spreadsheets that make extensive use of these capabilities are hard to maintain. What usually happens is the developer who originally built the spreadsheet will get called every time there is a problem or a change is needed. I have met developers who spent their entire careers developing and subsequently maintaining a single system. Some might call this job security; I call it boring.

- One of my biggest complaints, especially with Array Formulas, is they make the spreadsheet hard to debug. A powerful technique when developing a complex formula is to enter an initial version of the formula, apply it to all the rows in the table, and look at the specific rows where error conditions or other unexpected results occur. This is also an effective way to identify boundary conditions that may have been overlooked. The recreant formula can be corrected or enhanced <u>on that specific row</u> to ensure the change addresses the problem. Then the enhanced formula can be copied to all the rows in the table. The process repeats until all issues have been identified and addressed. This isn't possible with an Array Formula. Only the "master formula" (for want of a better term) can be modified, which means this technique can't be used. Good luck if you are trying to build something really complex.

- Compatibility is another consideration. Not all of Excel's advanced features are available in other spreadsheet implementations (such as earlier versions of Excel, Excel web, Excel mobile, Google Sheets, Apple Numbers, etc.).

- These features don't provide any capabilities that can't be accomplished, probably quicker and easier, in other more direct ways.

One of the weirdest (I can't think of a better way to describe it) and most frustrating issues I've encountered is that the value returned by some Excel functions[46] may or may not be an array depending how it is used! One such function that I use frequently is CELL.

The CELL function returns a variety of useful information about the cells in a spreadsheet. For example, CELL with the "filename" parameter returns a single value with the full file name (including path) and tab name of the specified cell. Other parameters return information about the type of data in the cell, how the cell is formatted, if the cell is protected, etc.

Recently I had occasion to create a spreadsheet that would be used to fill-in data on preprinted forms. Each tab in the spreadsheet supported one of the forms. Spacing was obviously critical to ensure the printing would be in the correct positions. To make sure that any subsequent maintenance did not alter the spacing I included CELL functions with the "width" parameter (which returns the column width) in each column below the bottom of the Print Area. Like any good spreadsheet developer, I put the formula in the first column and copied it to the right across the width of the print range.

Sorry – can't do that. When using the "width" parameter the CELL function returns a one-row, two-column array.

---

[46] Well, at least one that I've encountered.

The first cell in the array contains the column width, and the second cell contains a logical indicating if the width is the default width (TRUE) or if it was manually adjusted (FALSE).

While this second piece of information might be useful, I don't understand why the demons who labor in the dungeons of the Excel laboratory opted to return a two-column array rather than a two-row array. Since the function is intended to return the width of a column, doesn't it seem reasonable that someone might want to use it across multiple columns? Maybe whoever (or whatever) wrote that piece of code was in a bad mood that day.

Be that as it may, you can't put them in consecutive columns on the same row. You have to stagger them. The illustration below is how I had to deal with it.

FIGURE 14. USING THE CELL "WIDTH" PARAMETER

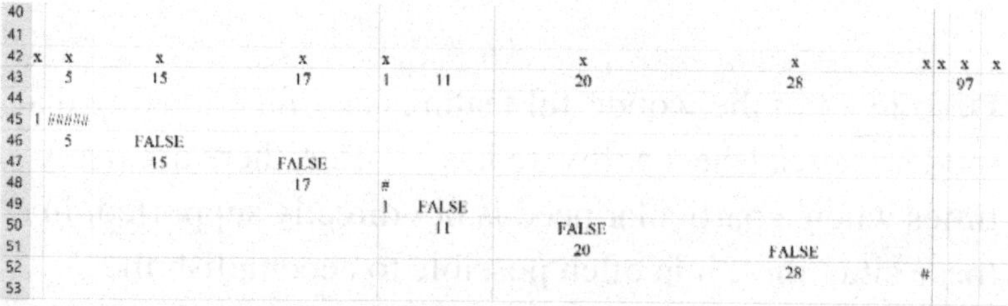

It is no accident that the examples in the Excel help are extremely simplistic. They are also most likely amateur and "single use" applications – the developer keys in a few rows of data, applies a relatively simple formula, gets a result and throws the spreadsheet away. These are probably the only types of problems where Tables and Array Functions can be used effectively. What about a production spreadsheet where a comparatively inexperienced user imports thousands of rows of data from another system and produces a complex report?

My experience over the years has been that the more complex the problem being addressed, the more likely any attempt to use these features will result in either obscure and difficult to find errors or more complexity than if the developer had simply used the basic spreadsheet constructs the way they were originally intended.

Special Situations

Despite all of the wonderful features and functionality that modern spreadsheet software has to offer, there are many times when a particular need is not directly supported. In these situations, it is often possible to accomplish the desired objective indirectly using the available capabilities.

We have already seen some examples of this, such as using a pair of lookup functions to access the fields in a table by name rather than by their position in the table. There are of course numerous other examples, some of which are presented below. I am not including these to provide you with a cookbook of cool spreadsheet hacks. You can do an internet search and find thousands of them. The lesson here is to understand the requirements of the problem you are attempting to solve, understand the capabilities the spreadsheet software offers, and learn how to apply them in innovative ways.

In other words, you need to think beyond the spreadsheet.

CASCADING PIVOTTABLES:

The PivotTable is one of Excel's most powerful features. A PivotTable can be used to automatically:

- Create a new table from data in another table.
- Consolidate / eliminate duplicates
- Sort the entries
- Create subtotals and grand totals
- Make coffee[47]

While PivotTables are an excellent mechanism for transforming data, they are not always ideal for presenting the results. I will discuss this shortcoming and ways to overcome it further in the next chapter. For now I will focus on their transformative capabilities.

A PivotTable uses a specially structured range of data (i.e. a table) as input and creates a structured list of data elements (i.e. a table) as output. With a little imagination, it is not difficult to realize that a PivotTable can be used as the input to another PivotTable. Or, in a more practical application, multiple PivotTables can be created, each displaying different data elements, and collectively used as input to another PivotTable.

Why would anyone want to do this? The simple answer is that the data required for a report might come from many different sources and at varying levels of detail. The PivotTable is the perfect tool for consolidating the different sources and organizing the data into a consistent format. From there it is easy to access the data and present it in a report.

The following is a real example of a real problem that I encountered at a major financial institution. A report was needed to show in a concise manner the current and projected future status of the projects in each of the major portfolios. The information needed for the report came from several different systems:

- Project Planning system

---

[47] Well, not yet. I think that feature is planned for a future release.

- Project Billing system
- Resource Planning system
- Capital Appropriation system

Each of these systems held information required for the report in different forms. What all the systems had in common was the Project ID and the Project Phase.

After the data from the different source systems was imported, scrubbed and organized, PivotTables were set up to collect each type of information. These "Source PivotTables" were all created on the same worksheet tab, arranged vertically one after another. The rows of each Source PivotTable were the Project ID, Project Phase, and the data in each was one specific type of information:

- (Sum of) Planned hours
- (Sum of) Actual hours
- (Count of) Resources assigned
- (Min of) Planned start date
- (Max of) Planned completion date
- (Sum of) Approved capital
- (Sum of) Approved expense
- (Sum of) Actual capital

- (Sum of) Actual expense

Another table was set up combining the data from all of these Source PivotTables. Each row in this "Combined Table" had the Project ID, Project Phase, type of information, and the data value. A "Presentation PivotTable" was inserted using the Combined Table as input. The rows of the Presentation PivotTable were Project ID and type of information; the columns were the Project Phase. Each section of the report simply had to do a lookup to get the required information in the appropriate cell in the Presentation PivotTable.

One caution though: When using this technique, it is necessary to perform a "Refresh All" command twice. Otherwise the final Presentation PivotTable won't reflect the updated data from the intermediate PivotTables.

SPORADIC DATA

A situation frequently encountered is that not every row of a table has every element necessary. The missing element could be part of the key used by a lookup or part of the source for a PivotTable.

In the PivotTable example discussed in the previous section, the Project ID would only appear in the first row of each Source PivotTable. For Excel to correctly build the Presentation PivotTable it must be in every row of the Combined Table.

This *sporadic data* problem is encountered frequently. Fortunately it is easy to address. It is exactly the reason that it was necessary to create the Combined Table as input to the Presentation PivotTable rather than creating it directly from the Source PivotTables.

To populate every row of the table, test the field being interrogated to see if it has data or is blank. If it has data, use it; if it is blank, use the field from the previous row. The exact test may vary depending on whether the field is expected to have an alphabetic or a numeric value. For example, if the field being interrogated is in column A, and the field being populated in in column Z, the formulas might look like this:

=IF(A15>0,A15,Z14) ← Use this if column A has numeric values
=IF(A15>"",A15,Z14) ← Use this if column A has text values

SEQUENCE NUMBERS

Another situation frequently encountered is the need to create a series of unique sequence numbers to identify the rows of a table. Ideally the formulas to create the sequence number should not require modification as rows are added or deleted. The problem becomes somewhat more challenging if the sequence number must be *hierarchical*. That is, instead of being a simple sequential number it also has significance.

A good example of this requirement is a project plan. A typical project plan might be organized as a series of Phases. Each phase is divided into several activities. Each activity is divided into several tasks. A number is assigned to each task that identifies the phase and activity it is part of.

It is easy enough to just assign a sequential number to every task, but this doesn't tell the user anything about the relationship between that task and the other tasks in the plan. Therefore, by convention, the task number is assigned hierarchically and includes the phase and activity. A commonly used system is as follows:

XX.YZZ

Where XX is the phase, Y is the activity within the phase, and ZZ is the tasks number within the activity. Obviously other conventions can be used, but conceptually they are all the same. The illustration below shows the upper left portion of what might be a typical task plan. The task number is in Column A. The phase, activity and task descriptions are in columns B, C, and D (respectively). For each new phase the whole number portion of the task number increases by one. For each new activity the tenths portion of the task number increases by .1. For each task the task number increases by .001:

FIGURE 15 ASSIGNING SEQUENCE NUMBERS

	A	B	C	D
1		Project Information		
2	Task	Project Description and Milestones		
3	Nbr	Phs/Act		Description
5	01.000	Business Initiative Support:		
6	01.100		Construction Projects (This Year)	
7	01.101			Remote Home Office
8	01.102			New Office - Columbus Acquisition
9	01.103			Home Office Relocation
10	01.104			Close Kirkwood Avenue Office
11	01.105			New Office - Johnson Street
12	01.106			New Office - Capital Mall
13	01.200		Construction Projects (Next Year)	
14	01.201			New Office - Toledo
15	01.202			New Office - Columbus
16	01.203			New Office - Charlotte
17	01.204			
18	02.000	Strategic Technology Projects:		
19	02.100		Phone System Upgrade	
20	02.101			Assess Requirements
21	02.102			Prepare RFP
22	02.103			Submit Capital Request
23	02.104			Select Vendor
24	02.105			Develop Deployment Plan
25	02.106			Deploy System
26	02.200		Business Continuity	
27	02.201			Assess Requirements

Having a unique task number is critical in a task plan. It uniquely identifies all of the requirements, assignments and status of the task, which can be used for such things as creating summary reports or establishing dependencies for subsequent tasks.

The challenge here is creating a formula to assign these task numbers. Obviously they could be assigned manually, but that would be inefficient and error-prone. The formula should work for every row, regardless of whether it is a phase, activity or task, and not have to be changed as rows are inserted or deleted.[48] As long as there is a consistent structure that the formula can be applied to then this can be done.[49] In the example shown, the phase header is always

[48] In a nutshell, this statement is the ideal for every formula.

[49] And this is the prerequisite for enabling this to happen!

in column B, and the Activity header is always in column C. The task descriptions are listed in column D. Not only does this make the task plan easy to read when it is presented, it provides the structure necessary to create a formula for assigning task numbers:

```
=
IF($B5>"",ROUND(MAX($A4:A4)+1,0)+N("Phase Header: add 1 to the phase number, activity and task = 0"),
IF($C5>"",ROUND($A4,1)+0.1+N("Activity Header: add .1 to the activity number, task=0"),
$A4+0.001))
```

As you can see, I have added new lines and comments to the formula for clarity. The formula is actually quite simple:

- If there is data in column B, then this is a phase header row. The phase number is incremented by one, and the activity and task are set to zero.
- If there is data in column C, then this is an activity header. The activity number is increased by .1, and the task number is set to zero.
- Otherwise this is a task, and the task number is increased by .001.

If rows are inserted or deleted the formula for determining the task number must be copied over the affected rows to correct the references. This is an easy and common process when updating and spreadsheet. The key point here is that no changes need to be made to the formula.

GENERIC SEARCH

Anyone who has used one of the major search engines is familiar with generic searches. As you type into the search bar a list of possible likely matches appears. This is an incredibly valuable capability, as it can guide your search and help you find the desired entry much quicker.

Because of its intrinsic organization and method of operation, a spreadsheet can't do a generic search *per se*. But there are ways to accomplish effectively the same result.

The first way is by using the Auto Filter feature. Select all the rows and columns in a table and activate Filtering. Filter drop-down buttons will appear at the top of the table range. Clicking on any of the buttons will display the list of the values in every row of the table. As you type in the Filter Search field the list is reduced as values are eliminated that don't match the search criteria. As multiple columns are filtered the list of valid values shrinks until you are left with a small number of rows meeting all of the specified criteria are visible.

This capability is extremely powerful and useful, but it has a significant shortcoming: The filtering criteria are applied in succession as AND conditions. In other words, applying filters to multiple columns in a table selects the rows that satisfy all of the conditions. For example, consider a spreadsheet that has a table (a really large table) of names, addresses, places of employment, age, etc. If you need to find anyone who lives in Ohio AND is over 60, Auto Filter is the perfect tool. If you need to find anyone who lives in Ohio OR is over 60, Auto Filter won't do it.

Fortunately there is an easy way to accomplish this. Add a column (probably at the right edge of the table) which I will call the Test column. Ensure Auto Filter is activated on this column. Use the blank row at the top of each column the table (remember the discussion on table headers?) as a place to enter search criteria for that column. I'll call this the Criteria row. Enter a formula in the Test column for each row of the table that checks each field of interest in that row against the corresponding value in the Criteria row using the FIND function. If the desired value is found in any of the columns the cell in the Test column returns a value of 1 (i.e. TRUE). For example, if row 4 of the table is the Criteria row, the formula in the Test column for row 26 of the table might look something like this:

```
=OR(
IF($A$4>"",IF(ISERROR(FIND($A$4,A26,1)),0,1),0),
IF($B$4>"",IF(ISERROR(FIND($B$4,B26,1)),0,1),0),
IF($C$4>"",IF(ISERROR(FIND($C$4,C26,1)),0,1),0),
etc.
)
```

For each column in the table that needs to be tested, the formula first tests the corresponding cell in the Criteria row to see if something was specified. If so, the FIND function will determine if the criteria value for that column appears in that row of the table. If not, FIND will return a #VALUE! error, and that segment of the formula will return a value of zero. If the criteria value is in that row, that segment of the formula will return a value of one. The OR will return a value of 1 if any of the criteria are met. Filtering on the Test column for a value of one will show only those rows for which any of the criteria have been met.

PARSING

Parsing is the process of extracting specific data fields from within a string. Sometimes the desired fields are located in fixed positions in the string. We saw examples of this in chapter 6. Sometimes the position of the fields, or even how many there are, may not be known. This is where parsing becomes a challenge.

Consider the following example: You have received an email. In addition to yourself, the email was addressed to dozens, or perhaps hundreds, of others. You would like to save all of the individual email addresses for future reference. Extracting the emails is easy – highlight the field in the email that has the addresses (TO: or CC:), copy, and paste it into a cell in your spreadsheet. The cell (we'll use B1) in the spreadsheet now contains a string, perhaps several thousand characters long, containing the email addresses. The challenge is to extract each address into a separate row in the table.

The success of any parsing activity depends on the string being parsed having some defined structure. The defined structure can be *positional* (i.e. the desired information is in a specific position in the string) or *delimited* (i.e. there is a specific character that separates the fields within the string). The parsing algorithm needs to know what the structure is to be able to extract the desired data.

In the case of a string of emails, the individual addresses are separated by a delimiter, usually either a comma or a semicolon. Extracting all the email address requires two formulas:

- The first formula (in cell A2) extracts the first email address by searching for the delimiting character, starting at the beginning of the string.

  =LEFT(B1,FIND(";",B1)-1)

- The second formula (in cell B2) copies the email string except for the first email address.

  =MID(B1,FIND(";",B1,1)+1,99999)

These two formulas are copied down as far as is needed to extract all of the email addresses. Each row will have a single email address in column A and the remainder of the email string in column B. Once all of the emails have been extracted the list (in column A) can be copied, and the values pasted to wherever required.

CROSS REFERENCING

As discussed in the "Basics" chapter, one of the underlying differences between spreadsheets and programs written in a traditional programming language is that data values are pulled into a cell rather than pushed by program code. Because of this, there is no way (without using a macro) to set a value in one part of a spreadsheet or workbook from a formula in a cell in another part of the spreadsheet or workbook. There is however a way to test that the required value has been set.

Consider the following: A manager has a spreadsheet to track the sales of his team. Each member of the team has a tab in the spreadsheet for tracking his or her results When a person joins the team a tab is created by copying a template to a new tab and renaming the copy of the template tab to the person's name. There is a Summary tab where the results for all the team members are totaled. The table in the Summary tab has the name of the team member, followed by various sales statistics. It is easy to put formulas in the Summary table that pull various values from the individual tabs. There is however no way for the Summary table to know that it has an entry for every person.

Instead, you can do the next best thing. In the template include a test to determine if the salesperson's name has been added to the Summary table. This can be accomplished using a VLOOKUP of the tab name (which can be retrieved using the CELL function with "Address" as the info type) in the Summary table. IF the VLOOKUP returns a #N/A error then the name is missing from the table. The formula can then display an obnoxious message reminding the new team member to add his or her name to the table in the Summary tab.

FIND THE LAST DAY OF THE MONTH

This is just for fun, as it is a good exercise in how to think about the data and use the spreadsheet capabilities to solve a specific problem. It is also a good example of the successive thought process that one needs to follow when solving a difficult problem.

Part of what makes Excel such a good tool is that it is constantly being upgraded to incorporate functions that are frequently used. One such function is EOMONTH. It was added to Excel in 2007 and provides the date value of the last day of the month. (Previously, this function was in the Excel Analysis ToolPak.) Prior to that, if you needed to know the last day of the month you had to find it yourself.

Finding the last day of the month is tricky, as there are a several variables in the data that must be considered:

- Months have a different number of days, ranging from 28 to 31.
- One month (February) can have 28 or 29 days depending on whether or not it is a leap year.

There are several possible approaches, and which is best depends on the needs of the problem and the data being used. One possibility is to create a table with the end-of-month date for each month for leap-years and non-leap-years. This has issues, because it isn't always easy to determine what year is a leap year[50].

If you are really masochistic or get paid by the hour you could imbed a call to an external data source to see if it is an election year, since in the United States one of the founding fathers with a really bizarre sense of humor thought this would be fun. I would recommend against this.

[50] The actual rule is: If the year number is evenly divisible by 4 it is a leap year, unless the year number is evenly divisible by 100 in which case it isn't, unless the year number is evenly divisible by 400 in which case it is. Now for the test: Was the year CE 900 a leap year or not? (Answer on page 306)

Fortunately there is a much simpler approach. Excel (and so far as I know all other spreadsheet software) stores dates as the number of days since January 1, 1900. Therefore, if a month has 31 days, then the value of the first day of the next month will be 31 more than the value of the first day of the current month, etc. Also, since the date is just a number, "day zero" of the next month is actually the value of the last day of the current month. So, we can determine the value of the end of the month as the value of the first day of the next month, minus 1.

Assume cell A1 has a date. Placing this formula in another cell will calculate the last day of the month:

```
=DATE(YEAR(A1),MONTH(A1)+1,1)-1
```

Technically, this shouldn't work. If the date in A1 is in December, then the "MONTH(A1)+1" portion of the formula would yield month 13 (not really a legal month number). However, in its never-ending struggle to be accommodating, Excel does exactly what it believes you wanted it to do and returns a date of December 31. Other spreadsheet programs might not be so forgiving, so we need to find something more robust.

Since no month has more than 31 days, adding 31 to the day should always yield a day in the next month. Therefore, if you add 31 to the day and then subtract the resulting day number this should give you day zero of the next month, or the last day of the current month. Something like this"

```
=A1+31-DAY(A1+31)
```

Unfortunately, this doesn't work either. If the date in cell A1 happens to be January 30 or 31, adding 31 days results in a date in March, not February. The formula will then return February 28 (or 29 if it is a leap year) as the last day of January.

A more rigorous if not aesthetically pleasing approach is to use the concept in the formula above but modify it so it will always work. This can be accomplished by backing the date to the first of the month before adding 31 days. (In the formula you actually have to add 32 days because when you subtract the day number of the current month you get day zero, not day one.) The modified formula will look like this

```
=(A1-Day(A1)+32)-DAY(A1-Day(A1)+32)
```

No, it's not as pretty. but it works.[51]

[51] How do I know it works? I constructed a simple spreadsheet where I tested 10,000 dates(I figured that was a pretty good sample. I calculated the last day of the month using this formula and the EOMONTH function and took the difference. On every row the difference was zero.

8. PRESENTATION: HAVE IT YOUR WAY

Claude Shannon (1916-2001) has been called the father of Information Theory. Among his many other accomplishments, in 1948 he published a paper called "A Mathematical Theory of Communication". In it he made many profound observations, one of which has particular relevance to this day: For communication to be effective, the sender must communicate in the language of the receiver.

In 1973 NASA launched Skylab, the first multi-national space station. Skylab was inhabited by both US astronauts and Russian cosmonauts[52]. One of the factors that made the Skylab mission successful was their adoption of Shannon's dictum: the Americans always spoke to the Russians in Russian, and the Russians always spoke to the Americans in English.

In some respects the amazing success of PC-based office productivity software such as electronic spreadsheets was a direct consequence of the fact that for the first time users, not the IT professionals, could get information in a manner that made sense to them. Before that, the way information ended up on reports was dictated by the capabilities and comprehension (or lack thereof) of the IT developers.

---

[52] They were astronauts, but the Russians had to come up with a clever name to maintain their pretense of superiority.

With the advent of spreadsheet software users could not only manipulate their data, they could also present it the way that they wanted to see it. And guess what – they liked it! In fact they became addicted to it. As the tools got more and more powerful, and the users got better at using them, their appetite for seeing the data their own way became an insatiable obsession. And this obsession has continued to this day – not only with spreadsheets, but with a myriad of follow-on presentation tools that have come on the market.

Beauty May Be Only Skin Deep, But….

Human eyes are pretty amazing, but we can't see the bits and bytes crawling around inside a computer as it does its magic. For that reason, presentation has always been an essential element of computing. In the earliest days presentation was nothing more than the blinking lights on a display panel (yes, I remember that). The ability to spit numbers onto a piece of paper was a major advance.

In the early days of the

MAINFRAME

this was limited to letters and text on 8 ½ X 14 greenbar paper[53]. The biggest issue though was (and still is) that the mainframe is predominantly the domain of the IT organization, which leads to the inevitable communication issues. The Agile approach to software development has attempted to address this, with limited success.

The advent of the PC and the software that can run on it created an entirely new paradigm. Though much less "powerful" than the mainframe, it offered the user something the mainframe couldn't: interactivity and virtually limitless personalization.

Over the years, in addition to their enhanced data transformation capabilities, the presentation capabilities of

---

[53] It was called greenbar because it had alternating lines of gray and light green. That is so long ago that now "greenbar" comes up in Word as a spelling mistake.

spreadsheet software have become ever more powerful. Today, Excel and its brethren boast an array of features whose sole purpose is to make the visual aspects of what the user sees on the screen or on the page more meaningful and appealing. Instead of just displaying numbers and letters (which is about all the original spreadsheet programs could do), spreadsheet reports[54] now include text and a wide array of graphics. Instead of just black and white, they now come in color. Instead of a single font there are hundreds.

Beauty may only be skin deep, but it is the first, if not the only, thing that we see when we look at something for the first time. As the old saying goes: There is never a second chance to make a good first impression. If the presentation doesn't engage and excite the audience, whether that audience is a group of accountants or the board of directors, the battle is already lost.

There are two absolute critical success factors when creating a spreadsheet report (or any report for that matter): accuracy and aesthetics. Accuracy goes without saying, but as discussed in the last chapter, an important element of accuracy is a common understanding of the meaning of the information being presented.

Aesthetics are more subtle, and are largely dependent on the expectations of the intended audience. They can also depend on the manner in which the information is being

---

[54] As with the term "file" used earlier, I am using the term "report" in the completely generic sense. A report is any construct in any format on any media whose primary purpose is to convey computer information to a human audience.

presented: What works on paper might not workon a tablet a cell phone screen.

There's an old saying that a picture is worth a thousand words. Personally, I feel that whoever came up with this saying did a great disservice to pictures: A picture is worth much more than a thousand words. And there's a good chance that as you add more and more words the intended meaning becomes more and more obscure. Anyone who has ever read a legal document understands this.

When someone first looks at a report their brain sees a picture. It is only after that initial look that they begin to discern the individual elements, and only after that they begin to interpret the information being presented. That is why it is so critical that the image of the report (i.e. the picture) be as appealing as possible. If the audience is a busy executive he/she may well take a glance and move on to something else unless the image is compelling.

What makes a compelling image is of course dependent on numerous factors, not the least of which are the intended audience, cultural norms, and the planned presentation medium. This may sound obvious, and obviously each situation is unique, but it is amazing how often this simple rule is ignored. In this chapter I will present a number of basic presentation guidelines that I have found to be powerful and effective. Not everyone will apply to every situation, but they should provide a good place to start. Following these guidelines won't guarantee a winning presentation, but ignoring them might guarantee a loser. Think about it.

An Embarrassment of Riches

In systems, as in life, one frequently encounters two different situations:

- Problems that have no obvious solution
- Problems that have too many solutions

Problems that are addressed with spreadsheet software often fall into the latter category. The software provides so many capabilities and so many options that novice spreadsheet developers often get carried away and end up creating reports that will probably accomplish little more than to give the user of the report a headache. Finding the right balance between boring and overwhelming is something of an art and usually takes experience to master.

PICTURE PERFECT

Always remember that a report is first perceived by the human brain as a picture. Therefore the developer must assume the role of an artist. As such, developing a spreadsheet report should be approached in much the same way as creating any other work of art. To create a picture that is both effective and engaging there are several factors to consider:

- **Message** – Yes, the first impression of a report is that of a picture, but it is a picture with a purpose. Why was the spreadsheet developed in the first place? What information should the report convey? What will the recipient of the report know after viewing the report that he or she didn't know before? These questions must be answered clearly and succinctly for the report to achieve its intended purpose.

- **Audience** – Often the same information can be presented in completely different forms depending on who is the intended audience. By extension, sometimes the intended audience will dictate the form. Columns of numbers may be appropriate for a working group session. A presentation for a board meeting will probably demand a PowerPoint slide with a summary[55].

- **Medium** – The medium in which the report will be presented has a significant impact on the design of the report. If the report is to be printed, the choice of paper size (letter, legal, 11X17, etc.) will determine how much information the report can contain. The same considerations hold if the report is to be delivered electronically, perhaps as a PDF sent as an email attachment. In this situation the preferred device of the recipients may need to be considered. What works on a PC screen may not work on a cell phone. If the developer is especially studious he or she might create multiple reports optimized for each type of device. If the report is going to be viewed actively online (i.e. as a spreadsheet containing one or more report tabs that is going to be sent to the recipients) it might be wise to include an index with hyperlinks to take the recipient to each report. Conversely, hyperlinks don't usually work on paper.

---

[55] Note: I can state from experience that it is always a good idea to have the supporting details handy in case there are questions!

- **Organization** – This is where the fun starts. This is where the developer must decide what the "picture" is going to look like. What will be the most prominent feature? Where should it appear? Will it be a graph or a table? The organization of the report should convey the message in the best possible way.
  As a (very) general rule, if the principal message of the report is a pattern or trend in the data then a graph may be the best approach. A graph already <u>is</u> a picture, and the viewer can absorb the message quickly with little mental effort. Graphs aren't always the best vehicle to create meaningful visual impact. If the data elements don't have an obvious or meaningful relationship a graph will probably only confuse.
  Each report is different. There are no hard rules when it comes to laying out a report (though there may be organizational standards that are expected to be followed). Developers must use their experience, sense of aesthetics, and understanding of the data to come up with the best way to convey the message. There are however some common and generally accepted principles that have proven to be helpful when followed:
  - **Overcrowding**. The single most common mistake I have seen in the spreadsheet reports I've reviewed is that the person creating it has tried to cram too much information into a single report or on a single page. The creator of the report probably has an in-depth grasp of the information being presented. The audience maybe not so much. Typically, they can only absorb small amounts of data at a time. Give them a baby dose and have supporting detail available if needed.
  - **Logical Layout**. The individual data elements that are most important should be the easiest to find and the ones that the viewers eyes are

immediately drawn to. This includes such things as the column headers, totals, etc.

- **Column Widths and Sequence**. This is especially critical when producing tabular reports. Beyond accommodating the maximum value of the data in the column, column width conveys an implication of importance. Where the data in each column has equal significance the columns should be equally prominent, if for no other reason than this prevents the viewer from subconsciously seeking significance that really isn't there.
  Since English is read from left to right, the leftmost column should be the most important. If the report has an explicit or implied sequence the leftmost column should have the data that controls the sort. The next column should control the sub-sort, etc. If there are logical groupings of the data (for example, this year vs. last year) the columns containing the data for those years (budget, actual, discrepancy) should be grouped together with spacing between the groups.
- **Centered/Boxed Headers**. A column header serves two functions: It identifies the data in the column and it provides information about the position, spacing and other attributes of the column. Centering and boxing the column header shows unambiguously where the column is and what it contains.
- **Follow Generally Accepted Conventions**. Audiences often expect certain data elements to be presented in certain ways. Classic examples include Social Security Numbers (XXX-YY-ZZZZ), telephone numbers (+1 (AAA) BBB-CCCC), and local currency formats.

There are also numerous situation-specific conventions. Many organizations have formatting conventions for customer numbers, account numbers, product identifiers, locations, etc. Regardless of how the data is stored internally, it must be presented in the format expected by the audience. (Remember Claude Shannon?)
Some conventions are dependent on the medium. In reports designed to be viewed "interactively" (either on a PC or a mobile device), underlined blue text is usually interpreted as a hyperlink. In a paper report this formatting is probably inappropriate and confusing.

CONSISTENCY

Implicit in the guidelines described above is the concept of consistency. Consistency helps the users of a report quickly recognize and interpret the information. Their brains can skip the "what am I looking at?" stage and go directly to "what does it mean?".

Consistency implies repetition, and it does not just apply to the data in a single report. Consistency can apply across similar reports, or even a collection of reports created multiple times. Many organizations hold regular status meetings at which a "report book" is reviewed. The book consists of a collection of reports that are generally the same each period.

One manifestation of consistency in many organizations comes in the form of report standards. Just as they might have standards for table layouts and other parts of the spreadsheets, they may also have standards for report layouts. These typically include specifications for what information must appear in the report header and footer, the fonts and colors to be used, etc.

Within a report, it is important that similar data types are presented consistently. For example, if a report has multiple columns of like numbers they should all have the same format (alignment, decimals, use of commas, etc.). But the reverse is equally important. Where multiple columns have unlike data that difference must be made as obvious as possible. For example, a report might have one column that shows units and the next column shows dollars. These columns should use different formats to make it clear to the user that their meaning is very different.

FORMATTING

In the context of a spreadsheet, formatting can be defined as any action that impacts the appearance of a single cell. This differs from organization, which describes the arrangement of the individual cells and their relationship to each other.

Formatting has seen significant enhancements since the early releases of Excel. From the early releases, which could only spit out black letters and numbers, Excel now provides an extensive variety of options. There are no concrete rules for how to effectively use these options to create a compelling presentation. What works best will vary based on the situation. The critical success factor is remembering that the options are available, and not being afraid to use them.

The most important formatting elements are as follows:

- **Font** – Excel provides access to the fonts included with the operating system. Out of the box, this is over 350 different character sets for Windows. These provide a diverse range of shapes and styles and support for several languages besides English. In addition to the common letters, numbers and punctuation, there are several fonts that entirely consist of various symbols, pictures and line segments.
  The choice of font can dramatically affect the visual impact of a report. For example, sometimes it is effective to have a bolder font in the header to offset a lighter font in the body of the report. The visual impact can also be enhanced by modifying the font's style (with bolding, italics or underlining) or size (from tiny to huge).
  Just don't go overboard. Make sure that the choices you make aren't just for show but improve the report's ability to convey its message. Never forget that sometimes less is more.

- **Color** – In my opinion, the invention of the copy machine (by Xerox in 1959) and subsequently the printers based on the same technology probably set civilization back by several hundred years. The early copy machines and printers could only excrete black and white documents. In doing so, they took away the dimension of color. The tremendous benefits that these machines offered seduced us into accepting that limitation. Even though today color printers are quite common, most reports are still only black and white. It may take humanity another hundred years or so to break this unfortunate habit. If you happen to be color-blind this probably isn't a big issue. But most of us aren't. We get most of our information through our eyes; we should use all the capabilities they have to offer.

The use of color to quickly convey meaning has actually been around for a long time. It has been the custom to present positive numbers in black and negative numbers in red since long before there was spreadsheet software, and even before computers.

Today the palette extends far beyond black and red. There are in fact 16,777,216 possible values (represented by the hexadecimal codes X'000000' to X'FFFFFF') that can be applied to both the displayed cell value (Font Color) or the cell background (Fill Color). In addition, there are Fill Effect, Pattern Style and Pattern Color options that can be selected. The possibilities are mind-boggling!

Certain colors convey generally recognized or culturally specific significance. We are familiar with the meaning of the green / yellow / red of traffic signals. The inference of these colors is frequently incorporated into status reports and other similar documents – Green means the activity is on track; yellow means there is cause for concern; red signifies there is an issue that requires attention. These are often referred to as "heat maps".

Colors can also convey emotional information and influence us more easily than most other means. Soft pastel colors convey warmth; hard bold colors are cold. Choosing the right color can influence how the viewers feel about the material being presented.

- **Boxing** – No, this is not about the so-called sport that appeals to the more primitive instincts of certain members of society. This is about clarification and

elucidation.
Excel provides a number of facilities for drawing lines around cells. The lines can vary in thickness. They can be solid or dashed. They can be on a single edge of the cell or all of them or anywhere in between. They can be at the edge of one cell or several.
Putting lines or boxes on a report can accomplish any number of purposes. Lines between rows of data can highlight how the data is grouped. Putting a box around a cell (or multiple cells) in the header can show the user of the report exactly where certain elements are located, how related elements are grouped, and how wide the fields are. Even individual fields in the report with special significance can be highlighted with boxes.

- **Miscellaneous** – In addition to the items discussed above, the "Home" tab on the ribbon contains a variety of additional formatting capabilities. These include features for manipulating the alignment of the data in a cell, the angle at which the data is presented, what to do if the data in a cell exceeds the cell's width, how to display numeric data, etc. All of these can be useful in certain circumstances depending on the needs of the report. As with all of the formatting options, the key is understanding how to effectively use the available tools.

Excel also offers a variety of styles, which in effect are automated formatting shortcuts. These work reasonably well for simple spreadsheets and can be a time-saver for quick-and-dirty applications. These really aren't adequate for complex or "professional grade" reports. For these the developer needs to understand the requirements of the audience, have a good sense of what makes a compelling picture, and knowledge of the available tools for realizing the vision.

One final note – an entire cell doesn't have to have the same format. Certain types of formatting (bold, italics, underline, color, font size & type) can be different for individual words or characters in a text string. This doesn't work for numeric values – how they appear is controlled by the number format code associated with the cell.

Excel provides a host of capabilities for improving the visual appearance of a report. Use them judiciously.

CONDITIONAL FORMATTING

Conditional formatting is almost an Excel within Excel. Instead of providing the ability to manipulate[56] the data in a table, it provides the ability to use the data in each cell of the table to change the way that data is presented.

A limited form of conditional formatting that only allowed three "levels" was first introduced in Excel97. It didn't become a fully functional tool until the 2007 release. Today it is an extremely powerful capability for managing the appearance of a report or screen based on a wide variety of developer-defined conditions. These include several predefined rules that can be applied to the data as well as the ability to specify complex formulas. Multiple conditional formatting rules can also be applied to the same cell. The formatting applied to a cell can even depend on the contents of another cell anywhere in the spreadsheet. (Wow!)

[56] Oops. I promised not to use that word.

A trivial example of conditional formatting, that was available since the early versions of Excel, is in the number format codes: One of the standard number display options is for negative numbers to be displayed in red rather than black and / or to be enclosed in parentheses.

Conditional formatting is an extremely powerful tool and can add another whole dimension to a spreadsheet – if it is used correctly. Like any formatting capability, too much glitz only detracts from the glamour.

One caution to be aware of: Conditional formatting takes priority over regular formatting. This may or may not be a good thing, depending. When you copy data from a source cell to another the formatting goes along with the cell contents unless you explicitly exclude it (using 'paste special > values' or 'paste special > formulas'). It is important to consider what impact the formatting of both the source and the destination will have, especially if one or both are subject to conditional formatting rules.

Conditional formatting can be used to highlight numerous conditions and situations that can't done with regular formatting. As an example, a spreadsheet I developed to enter financial transactions included these rules:

- Highlight which cells are unlocked to allow entry (using the CELL("protect",<cell>) function)
- Highlight potential duplicate entries
- Highlight unknown accounts
- Highlight undefined or invalid classifications

- Highlight entries requiring explanation because amount is over a specified threshold.

All of these rules act on various collections of cells in a single table.

One final thought about conditional formatting: As mentioned above, conditional format rules can include (potentially complex) formulas. The question that the developer needs to ask is: Is a conditional format the right place for the formula?

Consider the following situation: A large data table includes several columns containing formulas. The application allows the user under certain conditions to override the formula in one of those columns and enter a value. A conditional format is used to highlight any cell where the formula has been overridden so the entry can be reviewed. If for example cell W93 has been overridden with a value, the formula

```
=ISNA(FORMULATEXT(W93))
```

will return TRUE. This formula can be used to drive the conditional format.

This might not be the best way to do this. Instead, it might be better to add a column with a flag to show whether the cell in that row was overridden. The flag can be used to drive the format and those rows can be selected using AutoFilter. If the table has many rows, this is probably preferable to scrolling through all the table entries looking for formatted cells.

There is no fixed rule as to which is the correct approach. Whether to put the formula in the conditional format or the table will depend on the needs of the specific application.

Special Situations

The report, however it is presented, is in a real sense the ultimate goal of every spreadsheet. All the data acquisition and transformation activities performed in the various tabs are ultimately developed for the sole purpose of providing a user with information. The success of the spreadsheet is measured by how effectively that information is presented.

Excel provides the developer with an incredible palette with which to create that presentation. Often the presentation is more than simply formatting the data in a table and displaying it. In the following sections I am going to present many common problems encountered when preparing data to be presented. As always, the purpose here is not to show you how to address these situations, but how to think about these situations so you can address them yourselves.

HORIZONTAL PERSPECTIVES

From the earliest days of the mainframe, reports presented the user information in tabular, multi-column formats. This tradition has continued into the era of the spreadsheet and persists even today. While there are many good reasons why this format is still so popular, they are still plagued with the same basic problem: the human eye doesn't faithfully track horizontally across a page. It will invariably wander up or down, thereby confusing the entries at one side of the page with unrelated entries on the other.

In the old days, this shortcoming was handled by the paper itself. The "greenbar" paper that was used had alternating lines of differing colors to guide the eyes and ensure that they stayed focused on the correct entries. In today's world of laser printers and plain white paper (or simply viewing the reports online), that crutch is no longer available – unless it is specifically provided by the developer.

There are two techniques that I have encountered and used myself that are particularly useful:

- The first harkens back to the early days of the computer, when reports were printed on greenbar paper. Even though these reports were (by today's standards) pretty dismal, the paper that they were printed on had one wonderful feature: alternating lines were either light green or light gray. Therefore, as the eye traveled across the 14 inch width of the page it was always able to stay focused on the same line.
  Today's reports are usually printed on plain white letter stock and don't have that advantage. Never fear though, it is easy to incorporate (if, of course, it is appropriate for the type of report). This is especially helpful for densely packed tabular reports with many rows and columns.
  The first step is to include, off to the far right of the sheet well beyond the print range, a column with a simple counter. For example, let's say the counter is in column ZZ. The formula in cell ZZ117 would look like this:

  ```
  =if(ZZ116,0,1)
  ```

  This formula is copied into every row of the table for the entire print range. The result is that the value of the cells in column ZZ will alternate between 0 and 1.
  The second step is to add a Conditional Format rule using a formula that simply tests the value in column ZZ

for each row of the report. Where the condition in ZZ is TRUE (i.e., equal to 1), apply a Fill format of whatever color works best for your report, be it gray, green, or purple.
The result will look something like this:

FIGURE 16. "GREENBAR" FORMATTING

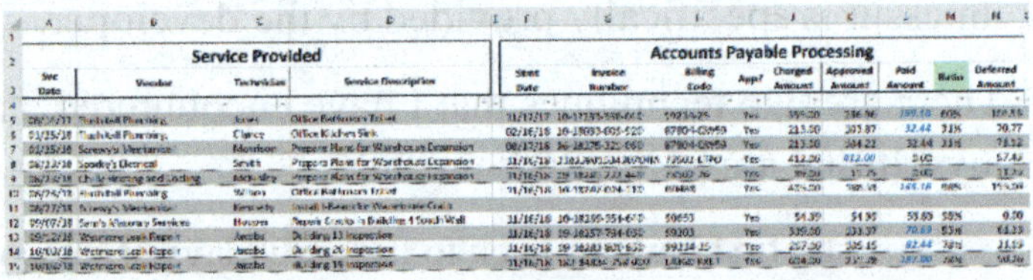

- The second approach is well suited to any report that has a "hierarchical" type of presentation (such as one that presents data by category and subcategory). This also takes advantage of the power provided by conditional formatting. If, for example, the category is in column A and the subcategory is in column B, create two conditional formatting rules. The first would, for example, put a heavy solid border above all the cells in the row if there is a value in column A. The second rule would do something similar, maybe a dotted border above all the cells in the row if there is a value in column B.

PIVOTTABLES DESERVE BETTER

PivotTables provide amazing power for organizing, accumulating and arranging the data for a report. But lets be honest – raw PivotTables are ugly. While this may be fine for analyzing the data, a PivotTable, with its Filters, fields headers, and dropdown boxes is not what you would want to show on a high-level presentation. Seriously, your audience deserves better.

Fortunately this is easily accomplished. Simply put nice-looking and appropriate headers at the top of the spreadsheet (rows 1 through whatever) above the top row of the pivot table. You can then hide the PivotTable control rows or, if the report is to be printed, just don't include them in the print range. Not only does this allow for better looking report headers, the column widths can be adjusted to best display the data without regard to the spacing needed by the PivotTable headers.

What if the PivotTable field headers change depending on the data? For example, the "columns" portion of the PivotTable are the 12 the most recent months. Each month when the report is run the list of months will be different. This type of situation can be handled by having the fields in the report header reference the fields in the PivotTable header. Or better yet, use the PivotTable column header as the lookup value in a table that has pretty headers and perhaps other variable information that will be displayed in the headers. For example, the column headers on a PivotTable might be the numbers one through twelve. This is done to force the PivotTable columns to display in ascending month sequence. Use the header value (i.e., the number) to lookup the month name (Jan, Feb, Mar, etc.) which will display as the header on the report.

FIGURE 17. WHICH DO YOU WANT TO SHOW YOUR USERS?

This one:

Valid?	1
Recent?	1
Event?	(All)
Bud Incl	1

Sum of Amount		Rec Mo												
Catg	Sub Catg	01.Dec	02.Jan	03.Feb	04.Mar	05.Apr	06.May	07.Jun	08.Jul	09.Aug	10.Sep	11.Oct	12.Nov	Grand Total
0 Revenue from Sales	1 Heavy Equipment	5,582	3,491	3,272	1,214	7,674	2,174	9,114	9,319	2,339	5,930	8,561	1,244	59,918
	2 Light Equipment	2,423	1,497	509	6,972	4,403	8,505	665	2,440	1,895	1,156	8,763	3,016	41,521
	3 Maintenance Supplies	7,057	7,231	8,531	2,874	1,107	7,626	6,327	4,574	9,796	7,362	6,285	6,028	69,997
1 Revenue from Services	1 Field Maintenance	2,216	2,209	8,219	6,654	8,091	1,458	9,528	3,867	5,757	1,680	8,614	2,610	60,904
	2 Rental Maintenances	9,428	8,898	8,166	1,370	8,902	251	8,954	166	1,704	6,548	2,681	2,395	59,263
	3 Consulting	2,674	8,873	6,630	2,822	7,154	5,356	7,900	2,626	9,573	2,654	6,488	5,705	68,456

Or this one:

	A	B	C	D	E	F	G	H	I	J	K	L	M	N	O
1	Classification					Category/Subcategory Rolling 12 Month Actuals									Rolling
2	Category	Subcategory	Dec	Jan	Feb	Mar	Apr	May	Jun	Jul	Aug	Sep	Oct	Nov	Actual
3		Grand Total - Revenue:	26,777	41,616	14,410	49,133	25,238	39,297	27,536	32,606	25,170	30,031	37,698	32,984	382,495
4		Expenses:	-4,890	-8,511	-9,760	-7,103	-7,396	-10,861	-32,356	-8,993	-12,218	-6,814	-7,455	-10,968	-127,326
5		Total:	21,887	33,105	4,650	42,029	17,841	28,436	-4,820	23,614	12,952	23,217	30,243	22,015	255,170
6															
14	0 Revenue from Sales	1 Heavy Equipment	8,974	5,831	2,707	8,779	5,345	5,870	2,422	7	5,548	3,933	7,604	7,583	62,601
15		2 Light Equipment	2,341	8,414	4,083	7,534	6,859	8,304	7,784	5,108	1,425	6,073	2,068	4,124	63,818
16		3 Maintenance Supplies	8,246	3,228	2,711	6,403	2,331	4,536	7,596	6,087	2,396	7,988	9,382	4,019	64,934
17	1 Revenue from Services	1 Field Maintenance	4,567	9,993	483	9,603	7,808	5,891	8,338	7,522	3,325	5,580	7,465	9,044	79,530
18		2 Rental Maintenances	823	8,858	4,031	7,708	2,485	5,284	1,167	6,926	2,613	3,682	3,168	4,858	51,619
19		3 Consulting	3,823	7,251	385	9,109	380	8,113	229	6,957	9,863	2,765	8,014	3,355	60,394

WHEN GRAND ISN'T

Previously we discussed why it is often desirable to have grand totals at the top of tables where they can be "frozen" and always visible as a user scrolls through the data. The same holds true for reports.

Traditionally, totals were placed at the bottom of the reports. In part this is because these reports could only be viewed on paper, and flipping to the end of the stack to see the totals was easy and convenient. Repeating the totals at the top of every page was a waste of paper. Also, the early software used to create the reports was limited and processed the data sequentially. Totals weren't available until the end when all the data had been processed. Thus traditions are born.

With spreadsheets this isn't a problem. All the data is accessible and can be used whenever and wherever it is needed. There is a different problem however: The desired grand total may not be just the sum of all the rows in the report. This will be the case if, for example, there are one or more levels of subtotals within the report. This can also occur if, for whatever reason, not every row of the report is supposed to be included in the grand total.

There are a number of ways this can be addressed. The two most common are:

- If there is a consistent number of subtotal levels the grand total can be obtained by summing all of the rows in the report and dividing by the number of levels. If every detail row is identified as being part of a district within a region, and there are district and region subtotals, the grand total is simply the sum of all the rows (including the district and region subtotal rows) and dividing by three.
  Note that this only works if the assignment of each detail record to a district and region is strictly hierarchical. If not, a different approach must be used.
  But – if the grand total is already available from some other source (perhaps it is inputted) it can be used as a control value to confirm that every row was properly assigned to a region and district.

- Since a report is also a table, it is possible to include fields that reference the fields in the report table outside of the print range (usually to the right). The header totals can then use these field to drive the grand totals. For example, a simple flag can be used to indicate that a row

of the table is a detail (vs. subtotal) row, and referenced in a SUMIF() statement in the header.
This approach will allow for more complex totaling as well. For example, if each row of the report is the activity for a line of business, the "flag" field can be the value of the line of business (or blank if the row is a subtotal row). Then there can be multiple SUMIF() statements in the header, one for each line of business.

If this discussion sounds a lot like the previous discussion about good practices when creating tables there is a good reason: The meat of a report is often just another table (with formatting). The same rules apply, and can provide a significant amount of clarification.

TURN THE PAGE (OR NOT)

One interesting requirement I encountered involved the following:

A report is required that displays the key metrics for active "critical / priority" projects (as identified by indicators in the project management system). The report would list the projects in assigned priority sequence, with grand totals at the bottom of the page for number of projects, costs, benefits, etc. The report will be incorporated into a slide deck at each month's executive committee meeting, and therefore formatting is critical.

Here's the catch: For the slide deck, no matter how many active critical / priority projects there are, only a specified number that will fit on one page will be included.

Furthermore, if the number of critical / priority projects takes less than a full page, the report must include blank lines so that the totals always appear at the same spot on the bottom of the slide. The committee members will use those blank lines to write in other projects they decide to elevate. A separate printed version of the full report (all active critical / priority projects) will be available on the server for review if needed. There are two lines of grand totals at the bottom of both versions of the report – one line for only the projects shown in the single page version and one line for all projects. (This is typical of the type of critical details that are often omitted from the original specifications and only become known during acceptance testing.)

The obvious solution is to simply create two copies of the report in the spreadsheet. Define the first report's print range so that it lists all the projects. Define the second report's print range to only print the first page.

Piece of cake!

But how do you get the nicely formatted grand totals at the bottom of the page on both versions of the report? And (oh, by the way) some projects may require multiple lines to print due to additional information that might not be present for every project. (See the comment about specification shortcomings, above.)

Ugh!

This is a situation where a pivot table won't really work as the report itself, but can be a powerful tool for collecting the data to go into the report.

The way we solved this was to create a report shell. The top of the shell had the formatted headers. The 500 rows below the header were formatted to include each of the possible lines (of which there were five) for one hundred projects (more than would ever be the number of active critical / priority projects[57]). Below that were 25 blank lines (the number of lines required to fill the single page version if there were no critical / priority projects[58]). Below that were the two grand total lines. The print range covered the entire shell from below the headers to the grand total lines. The header rows were set to repeat at the top of each print page.

With this arrangement we had every possible situation covered. A PivotTable was inserted on a separate tab that accumulated all of the relevant data for the active critical / priority projects. The PivotTable row definition included the project's absolute priority rank and other static information common to all projects (Id, title, responsible division, etc.) The columns definition included the totals for all of the variable project data.

In the projects section of the shell each field was populated with VLOOKUP() functions into the PivotTable using the project rank as key to access the relevant data. Beside (to the right) of the report shell was a series of columns with flag formulas to determine if the row had data for a project or not, the project's rank if it did, if any of the additional information was available and had to be shown, etc. This

---

[57] And yes, we had a control value that would flag an error if there ever <u>were</u> more than 100 active / critical projects. ☺

[58] And yes, there is a control value for the total number of lines on the one-page version. This, as it turns out, is used in later formulas.

information was used by formulas in other cells on the same row to calculate the starting and ending report row numbers for the project. Finally, a formula in the rightmost column (the "display control" column) used the data in these cells to set a switch (a value of zero or one) to determine if the row should be displayed based on the "mode" of the report, as determined by the value of the "mode switch" in row 2 of the display control column. Mode = 1 is single page; mode= 2 is full report.

In the blank rows section of the shell the formula in the display control column determined whether the row should be displayed based on how many blank lines are needed to fill the page (again, depending on the value of the mode switch).

The grand totals section of the shell was always displayed.

An AutoFilter was enabled for the entire print area and the display flags and control columns.

To create the reports after the data was loaded and processed, the users simply had to set the mode switch for the desired version of the report, refresh the AutoFilter, and print.

I have deliberately not gone into the details of the formulas in each of the cells in the report shell or the columns with the control values. These obviously will depend on the specific requirements of your reports. Suffice it to say the formula in each cell was actually pretty simple. The lesson here is to look at a problem and create a broad general solution that will meet the requirements in a broad set of circumstances. By combining the results from simple formulas one can solve complex problems.

Piece of cake!

DATA OVERBOARD!

Reports are often constrained by the available real estate of the media on which they are being created. Regardless of the media or the report format, an issue sometimes arises where there is more data than will fit in a particular field. Because of the limited spece available, it is often impractical to widen the column to accommodate the maximum width needed to display every field. If the adjacent columns are empty the data will spill over into the empty columns. This can ruin the appearance of the report.

This situation often occurs where data is being entered into a source table, and the table will then be printed (for verification or some other purpose). If the report table is being created by a PivotTable or some other means this won't be a problem because every field in every column will have a value of some kind.

There is an easy solution to prevent this from occurring: Where this occurs add a column on either side of the column that might overflow. (If the field that might overflow is left justified, add a field to the right. If the field is right justified, add a field to the left.) Put a space or other non-printable character(s) in each row of this extra column. This field doesn't have to be printed – the entire column can be hidden or its width set to zero. Since there is data in the field adjacent to the one that might overflow it won't overflow into the adjacent field, even if there is more data than will fit in the field.

This approach can also be used if the spreadsheet is being used to fill in a preprinted form. The empty areas between the form fields should have a space in them to prevent the possibility of one of the data fields overflowing past the end of the field on the form.

The important lesson here is that even though there are fields and columns <u>in</u> the report table, they don't necessarily have to appear <u>on</u> the report. In the next section I will discuss a situation where this makes it possible to create a simple report that would otherwise be very complex.

REPORTS WITH MULTIPLE PIVOTTABLES

Consider the following requirement:

A workbook contains a table with several years of detail transactions. In addition to its other data, each transaction has an assigned business unit and product line. A separate table has the budget values for each period for each business unit and product line.

Create a report that shows a summary of the transactions by business unit and product line showing the actuals, budget and difference (over/under budget) for the current period, previous period, and current period previous year. The report should be producible for any period covered by the data; "current period" will be determined by a supplied control value.

There are several challenges that must be addressed here. The first is how to identify the transaction records that need to be included in the report. This can be handled by adding three flag columns to the transaction data table:

- Curr Pd TY?
- Prev Pd TY?
- Curr Pd LY?

The cells in each column will have a formula that determines, based on the transaction date and the control value, if the transaction should be included.

Since the report has three sections, the simplest option is to create three PivotTables, each using one of the three flag columns as a filter. These three need to be combined to create the final report. It is a simple process to arrange the three PivotTables so they can collectively be used as the input to create the report.

Now for the wrinkle –there needs to be columns between the PivotTables with the actuals to lookup the budgets for each period, and another column to calculate the difference (over/under budget).

As discussed in the preceding section, the PivotTables can be placed in a report and the PivotTable control fields hidden so that they don't appear on the report. The problem here is, since there are three PivotTables, the rows on all three must align exactly. Otherwise what is displayed as data for one business unit and product line might actually be that for another. If, for example, the product line is new and there is no data for current period last year, that PivotTable won't have an entry for it, and everything from that line down will be off.

The solution is to include in the data table dummy entries for every business unit and product line. These entries will have zeros in all the amount fields (so they don't affect any of the totals), and ones in all the flag fields (so they get included in all of the PivotTables. In this way every PivotTable will always have an entry for every valid business unit and product line, and therefore all to the PivotTables will have the same number of rows.

BRIDGING THE INSIGHT ABYSS

The preceding sections have largely focused on situations encountered when creating "tabular" reports, though the thought processes described are applicable to any report type. As discussed at the beginning of this chapter, sometimes a picture is the best way to convey the information.

Excel provides several facilities for presenting a collection of data graphically. Excel calls these charts. but they really are pictures of the data. Being pictures, they allow the user to quickly understand the overall situation without having to examine the details of each individual data point.

In the following example I will demonstrate one way of presenting the data graphically, making it significantly easier to quickly identify the key areas requiring attention. This is also an example of how data can be included in the chart but not be displayed on the chart.

Figure 18, below, shows a portion of a typical small business financial statement. It shows the budgeted and actual sales, cost components, and net income for the period. Despite the fact that sales (both dollars and units sold) were better than expected, the net income for the period is significantly less than was budgeted. Why?

Figure 18b shows the contribution to the variance for each sales and cost component. These numbers are derived directly from the financial statement. While this shows each component's impact, the information is not easily discernable. Keep in mind this is a very simple example. There are more effective ways to present this information.

Figure 18c shows a Bridge Chart produced from the variance contribution data. In a bridge chart the elements being presented are arranged in order from left to right. The size of each bar reflects the size of the element. The position of each bar is set so that each bar begins where the previous bar ended. If the bars go up this indicates a positive contribution. If the bars go down the contribution is negative. In this example the leftmost bar is the budgeted income, the components are arranged in the order they appear on the financial statement, and the rightmost bar is the actual income.

FIGURE 18. PRESENTING DATA AS A BRIDGE REPORT

18a. Small Business Financial Report:

	A	B	C	D
1		Period Performance Data		
2		Performance Measure	Budget	Actual
3				
4				
5	Total:	Sales ($)	500,000	533,500
6				
7		Manufacturing Cost ($)	-225,000	-264,000
8		Transportation Cost ($)	-85,000	-132,000
9		Selling Expenses ($)	-40,000	-44,000
10		Overhead ($)	-40,000	-49,500
11				
12		Net Income ($)	$110,000.00	$44,000.00
13				
14		Variance (Actual - Budget)		-$66,000.00
15				
16		Units Sold (#)	5,000	5,500
17				
18	Average Unit:	Price ($)	$100.00	$97.00
19		Manufacturing Cost ($)	-$45.00	-$48.00
20		Transportation Cost ($)	-$17.00	-$24.00
21		Selling Expenses ($)	-$8.00	-$8.00
22		Fixed Overhead ($)	-$8.00	-$9.00

18b. Variance Contribution by Component:

H	I
Variance Contribution Due To	
Volume:	$48,500.00
Price:	-$15,000.00
Manufacturing Cost:	-$39,000.00
Transportation Cost:	-$47,000.00
Selling Expenses:	-$4,000.00
Overhead:	-$9,500.00
Total:	-$66,000.00

18c. Bridge Chart:

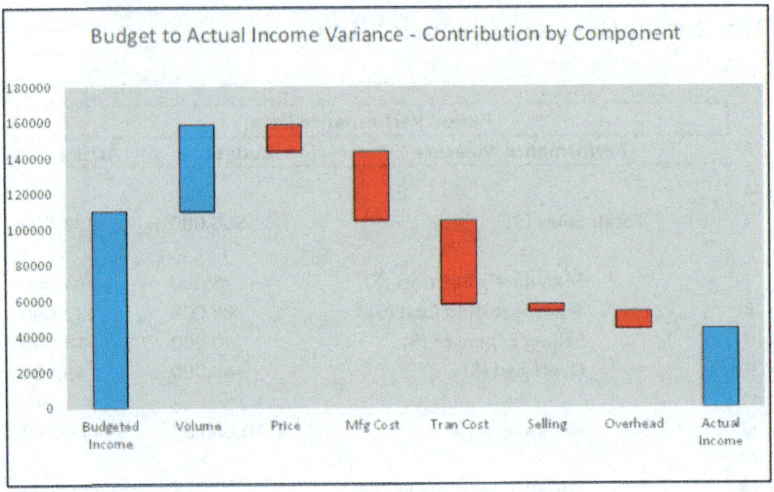

The Bridge Chart is created using a stacked bar chart. The first series in the chart is the height of the base of each bar. The height of the base is calculated so that the bottom (if the contribution is positive) or top (if the contribution is negative) equals the total height of the previous bar. The second and third series are the height of the current component's contribution (which is exactly the numbers shown in Figure 18b). The second series, which is blue, has the contribution amount if it is positive. Otherwise it is zero. The third series, which is red, has the contribution amount if it is negative. Otherwise it is zero. The effectiveness of the presentation comes from the fact that the first series is not displayed by making it the same color as the chart background. This makes each bar appear to be suspended, enhancing the "bridge" effect.

By presenting the information as a chart (i.e. picture) it is immediately evident that:

- Actual income is significantly below what had been budgeted for the period.

- Transportation costs were the most significant component causing this variance. Manufacturing costs were also a major contributor.

These insights are achieved quickly and easily without the need to even look at the numbers. Providing quick and easy insights should be the ultimate goal of every report

9. POLISHING

In the last three chapters we have examined the processes for accepting data from external sources, transforming the data into useful information, and effectively presenting the information. These three processes (each of which will often require multiple individual worksheets) embody the core elements required to make a spreadsheet an effective problem-solving tool. To be a <u>useful</u> problem-solving tool some additional elements are frequently needed.

Often there is information that is not part of the core functionality per se but are used by them. This supporting information can be single values or multiple values organized in tables. They can be relatively static or changed each time the spreadsheet is updated. How this supporting information is organized and handled can have a significant impact on the usefulness of the spreadsheet.

There can also be a need for information that is not used anywhere in the workbook. This information is provided solely to help the user understand what the spreadsheet does and how to use it.

People have been programming computers in their current form for a little over 70 years. In that time, organizations have developed many processes for managing the computers and the environment in which it operates. Some of these standards and procedures have become universally accepted and are used almost everywhere. Some are specific to and in response to the needs of each individual organization. Regardless of their type or origin, they are an essential part of the technology management process. Many organizations have learned, often through catastrophic failures, that without them the technology environment can rapidly deteriorates into chaos.

Creators and users of spreadsheets have largely tried to ignore the painful lessons of the past. There has been the mistaken belief that because of their ease of construction and apparent intuitiveness the old rules don't apply. After all, why impose a lot of useless rules that are only going to negatively impact the productivity of the people running the business?

As spreadsheets have proliferated and become more complex the situation has changed. It has become ever more important that they be maintainable and "provably accurate".

This focus of this chapter is two-fold:

- First, rules need to be defined (and followed) controlling the structure and organization of the spreadsheet itself.
- Second, rules need to be defined (and followed) governing the relationship between the spreadsheet, the user, and the environment in which it is used and maintained.

Tab Etiquette[59]

How the various elements of the workbook are identified and organized can significantly influence how efficiently the spreadsheet can be used and maintained. As mentioned earlier in the book, the users of a spreadsheet (especially in a large organization) are often not the same people who originally developed or currently maintain it. Therefore it is imperative that every part of the workbook be laid out and organized so it is as understandable as possible.

In Excel and most other spreadsheet software programs the individual worksheets are organized in tabs that appear across the top (or bottom, or side) of the screen. How these tabs are named and arranged can greatly aid (or hinder) one's understanding of how the entire workbook is organized and works.

INFORMATIVE TABS

Especially in a large, complex workbook, there should be tabs that aren't used in any formula and don't appear in any reports. These are the information tabs. Their sole purpose is to help the humans who interact with the workbook (i.e. the users, developers and maintainers).

---

[59] As I was writing this section the Coca Cola company announced that, after 60 years, they were discontinuing their Tab soft-drink line. In honor of this solemn occasion I am dedicating this section to Tab.

In general there are three types of information tabs. Depending on the size and complexity of the workbook it may contain some or all of them. Depending on the size and complexity the information may appear in a single tab or multiple tabs.

The three types of information tabs are:

- **Instructions** – This tab explains to the user the purpose of the spreadsheet and how to use it. It should at a minimum describe where and how data is entered, what control values need to be set, and the purpose of and how to produce the report(s).
  Ideally the instructions should be arranged step-wise, rather like a table of contents, so the user can follow them one after another to perform the necessary actions. As a convenience, it can be helpful to include hyperlinks to each of the tabs and cells that need to be accessed.

- **Explanation** – Where there are many individual worksheets or the transformation logic in certain parts of the workbook is particularly complex it can be helpful to have one or more explanation tabs. In essence this is the spreadsheet equivalent of program documentation. These tabs can include a diagram or flow chart showing how the worksheets relate to one another, explanations of particularly complex formulas, or any other information that would be useful to someone who needs to debug or maintain the workbook.
  (For this specific purpose I make exception to the general rule that spreadsheet software shouldn't be used as a word processor.)

- **History** – Large, complex workbooks that are part of an organization's production information processing and management reporting processes will often require periodic maintenance. Reference tables may need to be updated, transformation logic may change, new features may be added, etc. It is important to keep a record of these changes, both to facilitate the maintenance activities and to trace back to when and where a change was made if problems occur. The change history should include a description of the change, the tables affected, and information such as when the change was made, the version number, who made the change, etc.

CONTROL TABS

Control data can be defined as reference data that is supplied by the user of the spreadsheet at the time the spreadsheet is executed used to direct or modify the actions performed by the various elements of the spreadsheet. "Supplied by the user" means that this is data that is entered (as opposed to calculated). The data could be single values or entire tables. It could be keyed in or imported. But each cell of the reference data, however it is loaded into the spreadsheet, consists of values, not formulas. "Reference data" means that it is separate and distinct from transaction data. Reference data should not change regardless of how few or how many transactions are processed.

Control data can typically be divided into two distinct groups:

- Static Control Data, as its name implies, may not change at all from one execution of the spreadsheet to the next. These often consist of tables of reference data that are imported from some external source.

A good example of a static reference table in an accounting application would be the chart of accounts (COA). The COA is the list of valid accounting codes and related information. The COA would only change (and therefor need to be updated) when there is a change in accounting policy or practices. This may only occur once a year. The spreadsheet may be updated every day with the current accounting transactions to produce current financial reports, but the COA reference table wouldn't change.

- In contrast, the Dynamic Control Data would typically be updated every time the spreadsheet is used. Dynamic Control Data often consist of single values. Examples of this type of control are selection criteria such as dates used to determine the range of transactions to be included or the grand totals from external source systems (as discussed in chapter 7) used as a cross-check to ensure all transactions have been processed correctly.

If a spreadsheet has several small control tables it is probably more convenient to place them on a single tab rather than creating a separate tab for each. If multiple tables are placed on a single tab they should be staggered (both vertically and horizontally) so that changes made to one table (inserts or deletes to either rows or columns) don't inadvertently impact another table.

TAB ORDER

Having the correct tabs is important; having them arranged in a logical order is also important. In a large complex spreadsheet with many tabs (more than will conveniently fit across the bottom of the screen) a lot of time can be wasted looking for a particular tab.

Since we (English and romance-language speakers) read from left to right, it makes sense that the tabs should be arranged so that the most important / most frequently accessed tabs are on the left side of the tab list. In general, these will be the tabs that contain:

- Instructions
- Dynamic Control Data that is updated for each run
- Reports

The tabs where the data are inputted are also frequently accessed, so logically they should also be near the left. However, there is another consideration. It is useful to have the tabs arranged so that the data follows a logical flow from one tab to the next. In a sense data input is the beginning of the flow, and report creation is the end of the flow. As I just suggested that the report tabs should be placed near the left end of the tab list, so the data input tabs should be placed near the right (or at least to the right of the various data transformation tabs). This may seem awkward, since you must scroll through the intervening tabs to input the data. This annoyance can be easily circumvented by putting hyperlinks directly to the data input tabs in the Instructions. By extension, it is also useful to put a hyperlink back to the Instructions tab in the data input tabs (actually, in all the tabs) in a standard location such as cell A1.

Putting these considerations all together, this is my recommended order for the tabs (left to right):

- Instructions

- Dynamic Control Data
- Report Worksheets
- Transformation Worksheets
- Data Input (Transaction Data) Worksheets
- Static Control (Reference Data) Worksheets
- Explanation Data (Documentation)
- History Data

TAB NAMING

As you add tabs to a workbook Excel assigns them a generic name – Sheet1, Sheet2, Sheet3, etc. It is the responsibility of the developer to change the tab name to something meaningful. Some names are pretty obvious: The instructions tab should probably be named "Instructions". The report tabs should probably be identified with the names of the reports. The tabs where the raw data is inputted or imported might be named "Raw <data type>". The names for others, such as the transformation tabs, may not be so obvious but should be given names that are meaningful in the context of the spreadsheet.

The actual names chosen will vary based on the purpose of each individual worksheet and the conventions of the organization. Just be sure to change them. Subsequent users and maintainers of the spreadsheet will thank you.

TAB COLOR

This Excel feature doesn't get used nearly enough – the color of the tab can be changed. It can be extremely helpful for both users and developers if the different types of tabs (as discussed above) are each assigned a unique color. This makes the tabs easy to identify and helps prevent inadvertently going to the wrong tab for activities such as entering the raw data.

BUILDING A TAB CROSS-REFERENCE

In chapter 3 I discussed how spreadsheets present a radically different paradigm from traditional procedural programming languages. In particular, one key difference is that the cells in a spreadsheet "pull" data from other cells instead of "pushing" it to them. Because of this there are certain structures that cannot be created in a spreadsheet using native spreadsheet capabilities.

One such structure is an index or table of contents[60]. In the previous section I discussed structuring the Instructions tab as a Table of Contents with hyperlinks to each tab in the workbook. But how can you ensure that the instructions list all of the tabs, or that the names in the Instructions tab are

---

[60] If you do a web search for "Create an Index of tabs in Excel" you will find dozens of sites that will show you how to do that. These all involve creating a macro, which means the use of procedural program code. I stand by my statement.

valid tab names? The short answer is that you can't do this directly, but you can do the next best thing.

There are two aspects to the problem, and each must be approached separately:

- Ensure that all tabs are included in the Table of Contents. A cell in one tab can't push a value into a cell in another tab, but obviously a cell in the first tab can test the value of a cell in the other tab. To begin, define a name for the list of tabs in the Table of Contents. "TOC" might be a good choice. In a defined position of every tab (maybe cell A1 – it doesn't really matter) insert this formula:

```
=IF(ISNA(
        VLOOKUP(
            MID(CELL("filename",A1),
                FIND("]",CELL("filename",A1),1)+1,
                99),
            TOC,1,FALSE)
        ),"WARNING: TAB IS NOT IN THE TOC","")
```

  If the name of the tab containing this formula is in the Table of Contents (or more specifically in the table with the name TOC) the formula will return blanks. If not it will return the warning message.
  Here's how this works: The CELL function specifying the INFO_TYPE "filename" returns the complete path/filename/tab name where cell A1 is located. The file name portion is delimited with brackets ("[", "]"), so the data to the right of the close bracket is the tab name. The VLOOKUP function searches for the tab name returned by the CELL function in TOC. If the tab name isn't in TOC the VLOOKUP will return "#N/A" and the ISNA function will return a value of TRUE. The TRUE part of the IF statement will be executed and the warning message will be displayed.

Note that this only works if the developer includes this test in each worksheet, and takes the necessary action to add the tab name to the Table of Contents when the warning message appears. If not it becomes a management issue.

- Ensure that the Table of Contents only has valid tab names. What if the Table of Contents was set up with the expectation that several report tabs would be added to the workbook, but some of them haven't been? How would this be identified?
  This is different from the test performed above. In some ways it is easier to address since it doesn't require checking the value of a specific cell, only the presence or absence of a tab. For example, in a particular workbook the tab names portion of the Table of Contents is in column "C" and the "tab check test" is in column "H". To check if the tab whose name is in row 13 of the Table of Contents insert this formula in cell H13:

  ```
  =IF(ISERROR(INDIRECT(C13&"!A1")),
      "ERROR: MISSING TAB","OK")
  ```

  If the tab whose name is in cell C13 exists the formula will return "OK". If it doesn't, it will return the warning message. The beauty of this approach is that it is completely dynamic – No changes to the formulas are needed as tabs are added, removed, or renamed. It will always return the appropriate value based on whether a tab with the indicated name exists or not at the time.

The illustration below shows a portion of the Instructions tab for a moderately complex financial spreadsheet.

FIGURE 19. FINANCIAL SPREADSHEET INSTRUCTIONS TAB

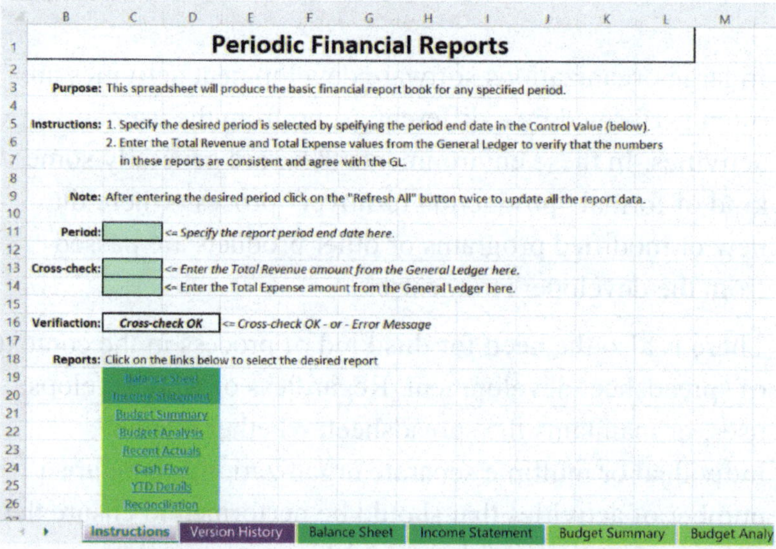

Ready for Production?

In large organizations software development activities are often performed by a different group from the user activities. In these environments there are generally some kind of formal "production turnover" process where the new or modified programs or other products are passed from the developers to the users.

There is also the need for this kind of process in the context of spreadsheet development. Regardless of who develops, uses, or maintains the spreadsheet, whether it is one individual or multiple separate organizations, there are a number of activities that should be performed to ensure the long-term integrity of the spreadsheet.

THE SYSTEM LIFECYCLE

Like traditional procedural programs (or actually any product) a spreadsheet will go through several distinct phases from its initial conception until it is ultimately retired. Depending on the size, complexity and function of the spreadsheet these phases may include any of the following:

- **Initial Development** – Somewhere, somehow, the initial idea for the spreadsheet pops into existence. It could be the result of someone's brainstorm or identified in response to a specific need. However it is conceived, what follows is a series of actions where the empty spreadsheet file is transformed.

- **Test** – At some point the spreadsheet is deemed ready and its accuracy must be verified. Data is loaded and the results are verified against some reference (a production report, calculated manually, whatever). If the results are not what is expected the cause of the discrepancy must be identified and adjustments made.
- **Quality Assurance** – If the spreadsheet is to serve a sufficiently important function (such as generating company management reports or other business-critical documents) there should be a formal quality assurance process. This generally consists of a defined series of rigorous tests performed by an independent QA team and may include loading and processing specific sets of inputs and the creation of a series of reports that can be verified. Equally important, it is the responsibility of the QA team to ensure that all applicable organization standards have been followed.
- **Training** – If the spreadsheet is used by someone(s) not involved in its creation or not familiar with its operation then training must be conducted to ensure that the spreadsheet is used correctly.
- **Production** – Once the spreadsheet is ready it can be used for production. Generally, this means that it is used to perform functions that directly support the operation of the organization. These typically include the loading of input data from some external source(s) (i.e. other than the spreadsheet) and the creation of "reports" in some format.
- **Maintenance** – Periodically spreadsheets must be updated to incorporate new features, accommodate changes in the business processes that they support, or changes in environmental factors (such as new versions of the spreadsheet software or the operating system). These changes may be made by the same group who use the spreadsheet or by someone else entirely. Regardless of who makes the change, the spreadsheet should go

through the Test / QA / Training phases again to ensure that the changes have been made correctly and the new functionality is understood.

- **Retirement** – At some point the spreadsheet will no longer be needed. This could be the result of changes to the business that render it irrelevant or being replaced by other capabilities.
- **Archive** – Once the spreadsheet is no longer needed it, along with all supporting documentation and other artifacts, should be placed in an archive. An archive should have two primary characteristics:
    - It should support permanent retention so that the files can be accessed if needed at some unspecified time in the future (for example, to support an audit).
    - It should be separate from and generally not accessible by the normal production environment. This is to prevent the files from inadvertent access, use or deletion.

Depending on the size and complexity of the organization where the spreadsheet is being developed and used, it may also be desirable to establish separate environments where the activities of each phase are performed. The defining characteristic of each environment is that there is a "wall" that prevents any action taken in one environment from impacting any of the artifacts (files, programs, etc.) in another environment. These walls are typically erected by the operating system security software.

The security software associates specific rights and permissions with each userid. Rights define the actions that the user is allowed to perform. For example, the rights associated with my userid allow me to execute the Microsoft Office Word and Excel programs. Permissions define what system resources (mainly files and folders) the user can access and what the user is allowed to do with them (create, read, update, delete). Security management, especially in large organizations, is a complex but essential activity. While it can sometimes be a source of frustration, security performs an essential function: It prevents us from inadvertently hurting ourselves.

PROMOTION

Regardless of whether the activities in each phase of the spreadsheet lifecycle are performed be one individual or multiple organizations, regardless of whether these activities are performed in one environment or several, there needs to be a process for "moving" the spreadsheet from one lifecycle phase/environment to the next. If nothing else, this is just good basic project management as it represents the completion of one phase and the initiation of the next. The larger the organization or the more complex the spreadsheet the more formal this process should be.

This process, regardless of how it is performed, is usually referred to as "promotion". It is well established for traditional programming efforts. It is often conducted by an external entity (i.e., one not directly involved in the project). This ensures that all the required activities have been performed and that the criteria for successful completion of the phase have been met. For the same reasons a formal promotion process needs to be implemented for spreadsheet development.

A typical promotion process for spreadsheets might look something like this: The "developer" builds the original spreadsheet or makes the needed changes to an existing spreadsheet. This is performed in the development environment. When completed the developer requests that the spreadsheet be promoted to the formal test environment. The "promoter" verifies that the required activities defined in the development phase of the project plan have been completed and moves the project artifacts to the test environment. The "testing team" is notified, and they conduct a series of tests. The process is repeated as each phase is completed until the spreadsheet is ready for production.

I have yet to find an organization that has implemented this entire process for spreadsheets. There is a simple reason why: it is cumbersome, time-consuming and incredibly annoying. [61]

Which takes us back to why users embraced spreadsheet software in the first place – so they wouldn't have to deal with all the bureaucratic BS that the traditional programming folks require. Ironically, in the early days of programming the programmers didn't deal with this either. But as systems grew larger and more complex, and they became critical components of the management processes, an ugly reality became apparent: mistakes can be devastatingly expensive. The investment in formal development methodologies (the proverbial ounce of

---

[61] I have also yet to find an organization that hasn't gotten bitten in the posterior at least once because they <u>didn't</u> implement this process!

prevention) is preferable to finding, fixing, and dealing with mistakes. In a classic example of history repeating itself spreadsheets are following the same evolutionary path.

The lesson is this: careful consideration must be given to the cost of implementing formal promotion procedures to ensure that mistakes aren't made against the cost of those mistakes when they occur. There is no one size fits all.

Standards, Guidelines & Conventions

The great power of spreadsheets is that they are both flexible and intuitive. As discussed throughout this book, this has helped make them the most widely used personal productivity application and an important management tool. But it has also created a false sense of complacency. The proliferation of spreadsheets has been a blessing, but if precautions aren't taken it will also be a curse.

There is an old saying: "*Freedom is not a license for chaos*."[62] Spreadsheets have an inherent structure (rows and columns), but there are virtually no constraints on what the developer can do within that structure. The result is that, without some guidelines, each new spreadsheet may bear no resemblance to the ones that preceded it.

Why is this a problem? Consider a typical situation: An organization has a number of individuals all of whom have developed a number of spreadsheets. These are updated on a regular basis to support a range of business activities.

One of the realities of life in the organization is that people move around. People come and go. They may move to different functions within the same group, or to a different group. People get sick or go on vacation, but the business doesn't stop. The activities supported by those spreadsheets must go on. The spreadsheets must be updated; the reports must be run.

If every spreadsheet in the organization has been designed and built differently it becomes difficult for someone new

---

[62] No, I don't know where this originated.

to update or use it. The more complex the spreadsheet the greater the difficulty. Best case, this difficulty is manifested as a severe impact on productivity. Worst case there is a very high probability of mistakes being made when the spreadsheets are updated and used.

There is another old saying: "*Familiarity breeds understanding.*"[63] The human brain likes patterns. We look for them as a shortcut so we don't have to start from scratch every time we encounter something new. If we can find a familiar box to use when we encounter something new it saves a lot of time and energy.

For spreadsheets this means that, at a macro level[64] they should follow a recognizable pattern. Though each serves a different function and contains unique logic the general organization and layout should be consistent.

To achieve the necessary level of consistency an organization needs to establish conventions for how various aspects of spreadsheet development and usage are to be performed. This is accomplished by defining a formal set of guidelines or standards. Their scope (applicable to an individual function or the entire organization) and what areas are covered will depend on the needs and maturity of each organization. The degree to which they must be adhered to can also vary. I use the following definitions: "guidelines" are recommendations or so-called best practices; "standards" are rules against which the product will be audited and evaluated. For larger organizations or

---

[63] As far a I know I just invented this one. Maybe it will become an old saying some day.

[64] Not that kind of macro. Think "grand scheme".

where the spreadsheet portfolio has high criticality it is better to use standards. This improves the odds that they will be followed.

Spreadsheet standards typically fall into two general categories:

- *Internal Standards*, which define the rules for the spreadsheet's design and organization.
- *External Standards*, which govern the physical spreadsheet file (in Excel terms, the workbook).

INTERNAL STANDARDS

Internal standards define the rules governing the spreadsheet's design an organization. They should cover the major elements of spreadsheet design and construction, and at a minimum should include:

- Overall structure, organization and design
- Required tabs, their contents and arrangement
- Table format and layout
- Range naming conventions
- Use of specific spreadsheet program features

This is by no means a comprehensive list. It should be considered a starting point for organizations that are developing spreadsheet standards. All the items listed above have been discussed earlier in this book. The message here is that organizations need to establish internal standards to ensure that their spreadsheets are designed, built and managed in a way that will minimize the potential for issues and errors. This will go a long way towards maximizing the productive use of spreadsheets as an effective business tool.

EXTERNAL STANDARDS

External standards define the rules governing the handling of the physical workbook files. For spreadsheets constructed using a recent version of Excel running in a Windows environment the files most frequently encountered have an extension of .xlsx (spreadsheet files), .xlsm (macro enabled spreadsheet files), and .csv. (comma separated variable files, frequently created or used when data is exported or downloaded from another system).

At a minimum, the external spreadsheet standards should govern the following:

- Spreadsheet file naming
- Spreadsheet file location

Note that I use the word "govern" here. Even more than the internal spreadsheet standards, the external standards must be defined and enforced. This is critically important if the integrity of the spreadsheets is to be maintained.

Spreadsheet File Naming Standards – Anywhere there are multiple spreadsheets in use, or a single spreadsheet is used multiple times, there should be a convention governing how the physical workbook file is named. The name should include some or all the following elements (not necessarily in this order):

- The organization that "owns" the spreadsheet
- A meaningful name describing the contents or purpose of the spreadsheet
- A version number of at least two digits
- If the spreadsheet data is updated multiple times the date of the latest update
- An indicator showing the status of the spreadsheet (e.g., development, test, QA, production, etc.)

Adherence to a rigorous workbook naming convention will go a long way towards preventing confusion and can be invaluable in situations where a problem needs to be researched or previous reports need to be audited.

Spreadsheet File Location Standards – The spreadsheet file location standards govern where the files will be saved. This is important for several reasons. The first, obviously, is so the file can be located if the current owner of the file is not available (for example out of the office or no longer employed). The second, and perhaps more important reason, is that access to the file can be controlled.

In most organizational environments there are file servers where production files are stored. Each department may have its own area on the servers which only it has access to. For example, the manufacturing department has an area where it maintains the production schedules, machine status, etc. The finance department has an area where the customer bills, invoices and ledgers are kept. You wouldn't want the two departments to be able to go in and access, alter or delete each other's files (either accidentally or deliberately). The operating systems used on the servers do an excellent job managing who can access what.

Another reason for having specified locations for the spreadsheet files is so that the promotion of the file through the different stages of the life cycle can be monitored and controlled. For example, if there is a specified Quality Assurance environment that can only be accessed by the person(s) who perform the QA testing, it is easy to see what spreadsheets are currently being tested. Also, typically, only the QA group or someone performing the QA function will have the authority to move a file into the production environment. This provides verification that the files in production have been reviewed by QA.

In larger or more mature organizations there will frequently be a training environment for new users or new spreadsheet applications. The training environment will have copies of the production spreadsheets and a predefined set of data files that can be used in the training exercises.

Once a spreadsheet has been developed updated and ready to be presented to the larger user community, the next challenge is keeping it that way.

PROTECTION

Protection ensures that actions on the spreadsheet can only be performed when, where, and by whom they are intended.

To accomplish this, protection must be addressed on three different levels:

- **Internal Spreadsheet Level** – Excel provides the ability to protect individual cells or entire worksheets in a variety of ways. Protection can prevent the user from changing, selecting, or even seeing the protected cells. This capability can be used to prevent errors caused by such things as accidently entering data into a cell that is supposed to contain a formula.[65] Use of Excel's internal protection capabilities is especially critical in situations where the user of the spreadsheet is different from the developer.

While the use of protection may be necessary and desirable it is important to recognize that it may prevent certain actions from being performed. For example, protecting selected cells in a worksheet may prevent the user from performing actions such as inserting or deleting rows, or updating pivot tables. The developer will need to test the spreadsheet after protection is enabled to ensure that the users can perform all the necessary actions.

Excel provides the ability to assign a password when activating security. Using a password will prevent the casual user from turning off the security and circumventing the protection. As with every application it is important to have procedures in place for assigning (and remembering!) the passwords[66].

- **File Level** – As with file naming standards, organizations need to have file protection standards. These standards will define which files need to be protected, and what type of protection is to be assigned. Passwords can be assigned to the physical workbook

---

[65] Excel actually gives a warning when an unprotected cell contains a formula.

[66] Bypassing an Excel worksheet protection password is relatively easy. I'm not going to explain it here, but a simple internet search will point you to lots of sites that will show you how. Nonetheless, it is still a good idea to use it. Many organizations have policies that expressly prohibit any attempt to circumvent security (Excel or any other application).

files. The option selected when the password is assigned determine whether a password is needed to update or even open the file[67].

- **Environment Level** – Environment level protection is closely linked to and typically part of the File Location Standards. Where the workbook file is located effectively determines who can access the file and what they can do with it. For example, if the Finance department has a server where all finance files are located, only persons with access to that server can access those files. In another example, a group that builds models which are used by many areas may put the model files on a server that they can update but other groups can only read. The other groups can copy a model file when it is needed, but they can't change the original "master" model file.

AND OH, BY THE WAY

It should be obvious, but it is worth pointing out because it is extremely important: Standards change over time. As new tools are added to the system portfolio, or changes are made in the system environment, the standards need to be updated to incorporate and reflect these changes. Therefore it is valuable to develop a protocol for tracking the different versions of the standards as they are updated and identify what major changes each includes. The documentation for

---

[67] Note that when a password is placed on the physical workbook file the contents of the file are encrypted. This can only be circumvented by using a password cracking tool or service to guess the password. If the password is "strong", this may be effectively impossible.

each spreadsheet should include the version number of the standard that was in effect when it was developed or last maintained. Each time a spreadsheet is maintained it should be checked to ensure that it conforms to the latest version of the standards.

One final note concerning standards. Standards can go a long way in ensuring that an organization's spreadsheet portfolio remains effective over an extended period. But they won't accomplish anything if they aren't followed. And they won't be followed if there isn't a mechanism to enforce them.

Let's face it, standards are a pain and a nuisance. Being inherently lazy, we (and I include myself in that group) aren't going to adhere to them if we don't have to. If I'm making a spreadsheet for my own use and no one else is ever going to see or touch it then I can do whatever I want. If it has issues or I pick it up some time later and can't figure out what I was doing at the time that's solely my problem.

Spreadsheets developed for use in an organization environment are inherently different. They may be developed, maintained, updated, or used by many different individuals. For this reason adherence to standards is a critical starting point for ensuring their long term usability and maintainability. And (it's worth repeating) because we are inherently lazy this won't happen if there isn't an enforcement mechanism of some kind.

The nature of the enforcement mechanism will vary depending on the needs, size and sophistication of the organization. If there is a formal Quality Assurance process adherence to applicable standards is a central component of their role. It could be the requirement that another member of the staff review the spreadsheet and sign-off before it is moved to the production environment. It could even be as simple as a dictate from the boss: "Thou shalt adhere to standards or you will be struck by lightening during your next performance review!".

How adherence to the standards is accomplished is less important than the universal consensus that they serve an important and valuable function. Adherence is essential, regardless of how they are enforced.

10. PROBLEM SOLVING

The purpose of a spreadsheet is to solve a problem.

It does this by presenting information to provide insights that cannot be easily ascertained simply by looking at the data. The problem to be solved might be some business-critical decision, or it might simply be idle curiosity.

Spreadsheets exist to solve problems.

In the preceding chapters I have discussed many considerations that help make spreadsheets effective problem-solving tools, and various practices that can help keep them useful. I have stated it before, but it is worth repeating: To build an effective spreadsheet the developer must –

- Understand the critical elements of the problem
- Understand the solution and how it relates to the available data
- Understand the capabilities of the available tools
- Understand how the solution must be presented so as to maximize it's effectiveness

In this chapter I present several examples of problems that I have encountered. In each example I discuss the key elements of the problem, and how spreadsheets were used to address them. Each example focuses on a different aspect of the problem-solving process and highlights the strengths and limitations of the spreadsheet as a problem-solving tool.

Presentation Power: Automated Merchandise Distribution

At one point in my career I worked at a large retail drugstore chain. The company had over 2,200 stores. Each store carried between 10,000 and 15,000 unique items (SKUs). In addition to the "regular" items that are always carried, the stores would at various times carry several hundred seasonal merchandise items.

Historically, each store received a single shipment of the seasonal merchandise which it displayed and hopefully sold over a multi-week promotional period. Unsold merchandise was returned to the distribution centers at the end of the season. The amount of merchandise allocated to each store was determined based on the size of the store and a lot of politics. Since the seasonal merchandise was very profitable, and there was no penalty for returning the excess, every store did whatever it could to get as much as possible.

The end result was that many stores ran out of merchandise early in the season and others had excess that had to be returned. In aggregate, between one third and one half of the seasonal merchandise went unsold and was returned at the end of the promotion period.

There had to be a better way.

We (the information technology organization) had recently completed a major project to implement scanning point of sale (POS) systems in all the stores. These systems provided valuable information about the activities in each store. Specifically, it was now possible to analyze the sales of each individual item in each store each day. We proposed using the POS data to calculate and send an initial "seed quantity" of the seasonal merchandise to each store and then replenish it as the merchandise was sold.

The calculations necessary to drive the distribution process were quite simple. The challenge was to prove to the senior executives that this was a significantly better approach. Given the sensitivity[68] of the subject the proof had to be compelling and simple. Simple because executives typically have an attention span that can barely handle a PowerPoint slide; they certainly wouldn't tolerate any massive report.

When the proposal was presented to the executive committee it was agreed that a test would be performed. Some of the items for the next seasonal promotion would be distributed using the automated allocation and the rest would be distributed using the current manual process. The results would be compared and presented.

The solution we developed was a one-page report that could be displayed for any seasonal item. The report was in the form of a grid. The horizontal axis showed the day of the promotion period (as fiscal week and day of the week). The vertical axis showed percent of merchandise sold, divided into 20 buckets (i.e., 00-05%, 05-10%, etc.). Each point in the grid showed the percent of stores that had cumulative sales of that percent of their merchandise by that day. The resulting pictures were stunning.

---

[68] "Sensitivity" is an organizational euphemism for "highly charged political nature".

The figures on the following pages show the results for two of the seasonal items. The two items shown are very equivalent (i.e., the only difference was their color), and therefor expected to have similar demand patterns. Both items had the same initial warehouse inventory to be distributed to the stores. Figure 20 shows the sales distribution pattern for the manually allocated item. Figure 21 shows the sales distribution pattern for the item that was allocated using the proposed automatic process. Both figures cover the first five weeks of the promotion period (the timeframe during which the items were appearing in the media advertisements).

A quick examination of Figure 20 reveals several problems with the manually allocated item. By the start of the fourth week 1% of the stores (approximately 20 stores) had already sold out. By the end of the fifth week this had risen to 4% (approximately 90 stores). Furthermore, by the end of the fifth week 1% of the stores had sold almost none of their allocation. The sales for the rest of the stores were almost uniformly distributed across all the percentages.

By contrast, as can be seen in Figure 21, the sales pattern for the automatically allocated item was significantly better. By the end of the fifth week none of the stores were sold out, and every store had sold at least 30% of the merchandise they had received. The overall sales distribution pattern for all the stores was very tight, meaning the merchandise was being sent to the stores where it was selling.

At the end of the promotion period less than 5% of the automatically distributed merchandise had to be returned to the warehouse, compared to 35% for the manually distributed items.

Building this spreadsheet was actually quite simple. The data input aspect of the analysis consisted of importing two files: Sales data by item by store by day was readily available from the POS system. Merchandise shipments by item by store by day was obtained from the warehouse management system. The transformation aspect consisted of creating tables to accumulate sales and receipts by store by day and dividing by cumulative shipments to the stores to calculate a percentage. A pivot table was used to count the stores that fell into each bucket. Dividing each bucket by the total number of stores yielded the percentage.

The challenge was how to present the several-hundred-thousand data points in a manner that would convince the executives. This was the critical element of the problem. Once we understood how to do this the rest was easy.

When we presented the results of the test we set up three PCs connected to large display screens. On two of them we showed side-by-side comparisons of similar items. The third was used to present supporting detail if requested.

It was very satisfying watching the faces of the executives in the room as they grasped the meaning of the information we were showing them. Some of them saw it right away; others took a while. By the end of the meeting everyone agreed that automated allocation was the best approach.

The success of the project depended entirely on an innovative presentation of the information.

Figure 20. Manual Allocation

Seasonal Merchandise Distribution Analysis - Item #130870 (Manual Allocation)

Cumulative Percent of Each Store's Allocation Sold by Day During Promotion Window

%Sold	Percent of Stores																																		
95-100																							1	1	1	1	2	2	2	2	2	3	3	4	4
90-95																							1	1	1	1	1	1	1	2	2	2	2	3	3
85-90																						1	1	1	1	1	1	1	2	2	1	1	2	2	2
80-85																					1	1	1	1	1	1	1	2	1	1	2	2	2	2	2
75-80																				1	1	1	1	1	1	1	1	1	2	2	2	3	2	3	4
70-75																			1	1	1	1	1	2	1	1	2	2	3	3	3	2	3	4	4
65-70																			1	1	1	1	2	2	2	2	2	2	2	3	3	3	4	3	4
60-65																		1	2	2	2	1	2	2	2	2	2	3	4	4	4	4	4	3	3
55-60															1	1	2	2	2	2	2	2	2	3	3	3	4	4	4	4	5	5	5	5	5
50-55														1	1	1	1	1	1	1	1	1	2	3	3	4	4	4	4	4	4	4	4	5	5
45-50											1	1	2	2	2	2	2	2	2	2	3	4	4	4	5	4	5	4	5	6	6	6	8	7	7
40-45										1	1	1	2	3	3	3	4	4	4	5	5	5	6	6	6	7	7	8	8	8	8	9	8	8	9
35-40									2	1	2	3	4	4	4	4	5	6	6	6	7	8	9	9	9	9	9	9	10	9	10	9	10	9	9
30-35							1	1	2	3	3	5	5	5	7	7	8	9	10	11	10	10	10	10	10	10	10	11	10	11	10	9	9	10	9
25-30					1	1	1	2	3	4	4	5	6	8	8	9	9	9	10	10	10	9	10	11	10	11	11	9	8	8	8	8	7	7	7
20-25			1	1	1	3	3	3	5	7	8	11	12	12	13	13	14	14	13	13	13	13	13	11	11	10	10	9	9	9	9	9	8	7	7
15-20		2	2	4	6	6	8	9	10	13	16	16	16	18	18	18	17	17	16	16	15	14	14	13	13	12	11	10	8	8	7	7	6	6	5
10-15	2	5	7	9	11	13	16	20	24	27	25	24	23	21	18	18	16	15	14	13	13	12	10	9	7	6	6	5	5	4	4	4	3	3	3
05-10	4	15	20	26	31	35	36	35	32	27	23	20	16	13	10	9	8	7	7	6	5	4	4	3	2	2	2	2	2	2	2	2	1	1	1
00-05	93	74	66	56	46	38	31	25	17	11	9	7	5	4	4	3	3	3	2	2	2	2	2	2	2	2	2	2	2	1	1	1	1	1	1
FWk:	38							39							40							41							42						
Day:	1	2	3	4	5	6	7	1	2	3	4	5	6	7	1	2	3	4	5	6	7	1	2	3	4	5	6	7	1	2	3	4	5	6	7

Note: *Percent of stores for each day may not total to 100% due to rounding.*

FIGURE 21. AUTOMATED ALLOCATION

Seasonal Merchandise Distribution Analysis - Item #130872 (Automated Allocation)

Cumulative Percent of Each Store's Allocation Sold by Day During Promotion Window

Percent of Stores

%Sold																																			
95-100																																			
90-95																																			2
85-90																																	2	4	4
80-85																																1	3	4	6
75-80															1	1	1	1				1	1	1								2	4	6	7
70-75													1	1	1	2	1	1	1	1	2	2	2	2	2	2	1	1	1	2	2	4	6	8	8
65-70												1	1	2	2	2	2	2	2	3	2	3	4	4	3	2	2	3	4	5	6	5	8	9	10
60-65											1	2	2	2	2	3	2	2	2	3	4	6	5	4	4	5	4	5	7	7	9	9	9	10	10
55-60											2	3	3	4	4	4	4	5	7	8	7	7	7	7	6	7	8	9	8	9	10	11	10	9	12
50-55										2	2	3	2	3	3	5	5	6	5	5	5	6	6	7	7	7	8	10	10	9	10	9	10	8	11
45-50										3	3	3	4	5	4	5	6	8	8	8	8	9	9	10	11	12	12	12	11	11	10	10	11	10	9
40-45									2	3	3	6	7	6	8	7	8	9	9	10	10	11	11	11	10	10	11	12	12	13	11	12	10	9	8
35-40									2	3	4	7	7	8	9	9	9	10	10	11	11	11	11	13	13	13	12	12	12	13	10	8	8	5	6
30-35			1	1	1	1	1	2	6	10	12	11	11	12	13	13	13	13	14	14	13	12	12	13	14	13	13	12	12	13	14	13	8	8	5
25-30			1	1	2	1	2	2	5	8	9	10	10	11	11	11	11	12	10	11	11	9	9	8	8	9	9	8	8	7	6	5	7	5	
20-25		2	2	2	2	4	5	6	10	13	12	11	12	12	11	11	12	11	10	9	9	8	8	8	9	7	7	7	6	6	5	5	2		
15-20	1	4	5	4	7	9	10	10	17	17	16	15	15	13	10	9	9	9	9	7	5	5	4	5	5	4	4	4	4	4	3	2			
10-15	5	8	9	13	14	17	20	22	22	17	16	12	10	9	8	6	5	4	4	3	3	3	3	2	2	2	1	1							
05-10	14	22	28	32	36	37	37	35	22	13	8	5	4	3	2	2	2	1	1	1															
00-05	77	60	51	43	34	27	21	18	8	3	2	1	1	1	1																				
FWk:	38							39							40							41							42						
Day:	1	2	3	4	5	6	7	1	2	3	4	5	6	7	1	2	3	4	5	6	7	1	2	3	4	5	6	7	1	2	3	4	5	6	7

Note: *Percent of stores for each day may not total to 100% due to rounding.*

Algorithms vs. Heuristics: Test Data

There are two general methodologies that can be used to solve a problem: algorithms and heuristics.

An algorithm can be defined as a set of rules or a sequence of steps that when followed produce the desired result or answer to a specific question. Algorithms are the heart and soul of computer programs. This is true regardless of whether the program was written in a procedural language (such as VBA) or a non-procedural one (such as Excel). In a procedural language program a defined sequence of instructions (steps) are executed in order unless the order is changed by testing for defined conditions (rules). In a spreadsheet the "steps" and "rules" are the formulas contained in each cell. The "sequence" is determined by the cell reference hierarchy[69]. Another characteristic of algorithms is that, for a given set of inputs and environmental factors, they always yield a unique solution.

Humans aren't very good at carefully following a defined sequence of steps or set of rules. Instead, most of the time we use a heuristic approach. Heuristics involve the use of multiple, often vaguely defined, criteria to evaluate a collection of options or alternatives. Almost everything we do is driven by heuristic problem solving. When we sit

---

[69] Excel keeps track of the cell reference hierarchy internally so it knows which "downstream" cells must be reevaluated whenever a given cell is modified. It also uses this to identify "circular references" (e.g., the formula in cell A1 refers to cell B1; the formula in cell B1 refers to cell A1.) which cannot be evaluated. Excel can display this hierarchy graphically by invoking the "Display Dependents" or "Display Precedents" features.

down at a restaurant and look at the menu we don't use an algorithm to decide what to eat ("Let's maximize the value we get by selecting the entrée with the highest ratio of calories to price."). Instead we use a heuristic approach ("What looks good to you?").

Spreadsheets are very good tools for implementing algorithmic solutions to many problems. Sometimes however there are factors to be considered that cannot easily be incorporated. Or the solution might require successive attempts to narrow the range of possibilities and settle on an optimal solution. Excel provides a "Goal Seek" functionality for situations like this. Sometimes however it's difficult to define exactly what constitutes the optimal solution. This is where the interactive nature of the spreadsheet software can be invaluable.

A project I was involved with at a large bank involved the implementation of a new type of monthly statement for the bank's investment clients. The new statements would show information specific to each type of asset held in each client's portfolio. A client might have a single asset type or dozens. We needed to ensure that the new programs would produce accurate and visually appealing statements for every type of asset a client might have.

The bank had approximately 15,000 investment clients at the time. Between them, they had 34 different asset types in their portfolios. The problem we faced was to choose a subset of clients to serve as a test set. We wanted the test set to be as small as possible to facilitate testing, but still cover every asset type. Because a large number of individuals would be reviewing the results of our tests we also wanted the ability to exclude any potential candidate "just because" (senior bank executives, local business leaders and politicians, etc.).

Computationally, identifying the subset of elements in a large set of possibilities that meet a specific criteria is extremely difficult[70]. We were not going to be able to write a program to give us the best test set. Instead we developed a spreadsheet to facilitate selecting a good set of test candidates manually.

The spreadsheet we constructed was quite simple yet proved effective for identifying a workable test set. Data from the client management system was imported and loaded into a table. Each row of the table contained the data for one client. The columns represented the 34 different asset types. Each cell in that part of the table had the number of assets of that type (i.e., unique CUSIPS) held by that client. For example, if a client had 14 different zero rate corporate bonds, the zero rate corporate bonds column in that row had the number 14.

The rows at the top of the table had headers and totals:

- Rows 1 & 2 – Headers
- Row 3 – Flag (0 or 1) to indicate the asset type still needs to be covered (i.e., that asset type is not held by any of the clients selected to be part of the test set). Cell E3 has the total of the flags in row 3.
- Row 4 – The total number of assets of each type held by the clients in the test set. Cell E4 has the total number of clients selected for the test set.

---

[70] "Extremely difficult" as in functionally impossible. The difficulty of this problem has led it to be used as the basis for certain public-key cryptographic systems.

- Row 5 – The total number of assets of each type held by all clients. Cell E5 has the minimum of totals in row 5.

The columns on the left side of the table had various indicators to support the test set selection process. Rows in the table where the indicators had specific values could be selected using auto-filter:

- Column A – Flag (0 or 1) to exclude this client from consideration.
- Column B – Count of asset types held by the client that are not already covered (i.e., not also held by previously selected clients).
- Column C – Flag (0 or 1) to indicate the client is a member of the group of clients with an asset type that is held by the minimum number of clients.
- Column D – Flag (0 or 1) to indicate client holds asset types that still need to be covered.
- Column E – Flag (0 or 1) to indicate client is selected as part of the test set.

Figure 22 at the end of this section shows the top left portion of the spreadsheet at the start of the test set selection process.

The selection process proceeded as follows:

- At the start of the process, Cell E5 has a value of 1, meaning there are asset types that are held by a single client. Column C has a value of 1 for each client who holds one of those asset types. Cell C7 show the number of flagged clients (6). Conditional formatting rules highlight the columns for those asset types.
- The rows for those clients are displayed using auto-filter. Since each of these clients hold an asset type that isn't held by any other client all six of them must be selected. But – those clients may also hold assets of other types as

well. Those asset types are now covered, so no further selections need to be made for them.

- After those 6 clients are selected Cell E5 has a new value (3, as it happens), and the rows for each client who holds those asset types are flagged. The rows for those clients are displayed using the auto-filter.
- Column B shows the number of not-yet-covered asset types also held by each client. The client holding the most not-yet-covered asset types is selected. (Note that this is suggested but not required. The developer creating the test set could select any of the 3.)
- Each time a client is selected the counters and flags are automatically updated (recalculated) and the process is repeated until all the asset types are covered.
- Where multiple clients meet the same criteria the person using the spreadsheet to create the test set uses their judgement to select the client that will provide the most meaningful results. For this reason, each time one of the developers created a test set they could end up with a different collection of clients (beyond the first 6, which appear in every test set).

Figure 23 shows the top left portion of the spreadsheet at the end of the test set selection process. Cell E4 shows that a test set of only 20 clients was adequate to cover all the asset types.

In all it only took about 10 minutes for the team to develop a very good test set that covered every asset type with relatively few clients. Developing a program to select an optimal test set would have taken a long time and (as mentioned above) would require a VERY long time to run.

The success of this approach depended on two factors:

- The use of a heuristic approach. Instead of trying to search the "best" solution using an algorithm the team settled for a "good" solution that worked.
- The interactive nature of the tool. The developer could select a candidate, see its impact, and keep it or try another.

One final note, and a caution to all spreadsheet developers: The PC provided to me by the bank was not very powerful. To develop and run the spreadsheet with this much data I had to copy the input files onto a flash drive and load them onto my much-more-capable personal laptop. The team then used my laptop for the test set selection runs.

This was a data security violation. The project manager and I met with our executive sponsor and had to explain the issue to her. She was impressed when I demonstrated the process and stated that this same approach could be helpful in several other areas as well. Shortly afterward the team received several new and upgraded PCs.

And they all lived happily ever after!

FIGURE 22. TEST SET SELECTION AT START

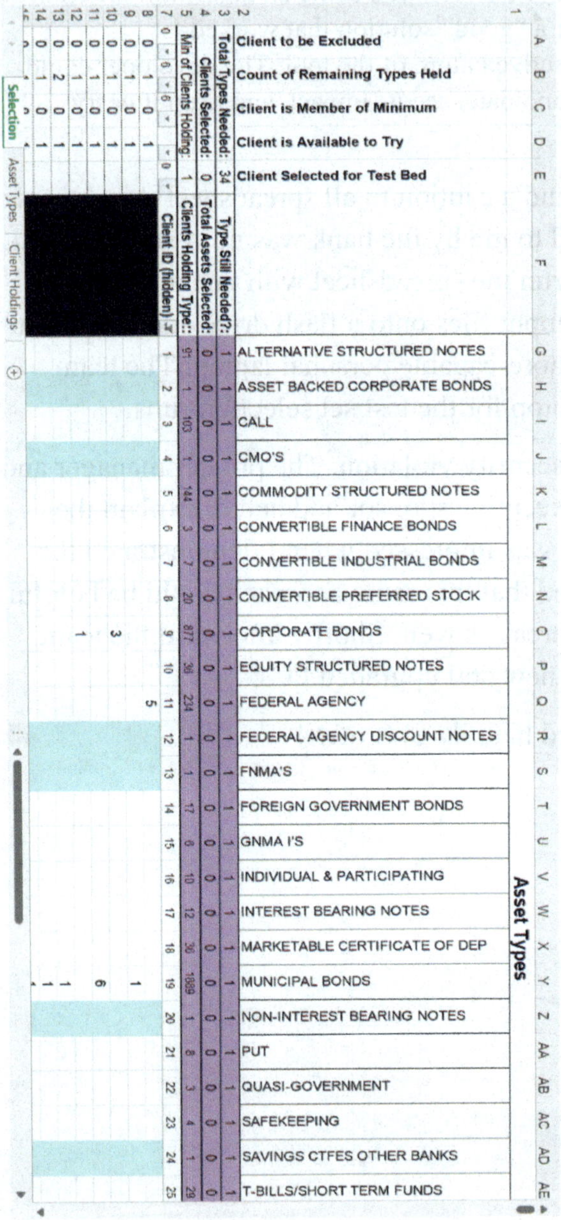

FIGURE 23. TEST SET SELECTION UPON COMPLETION

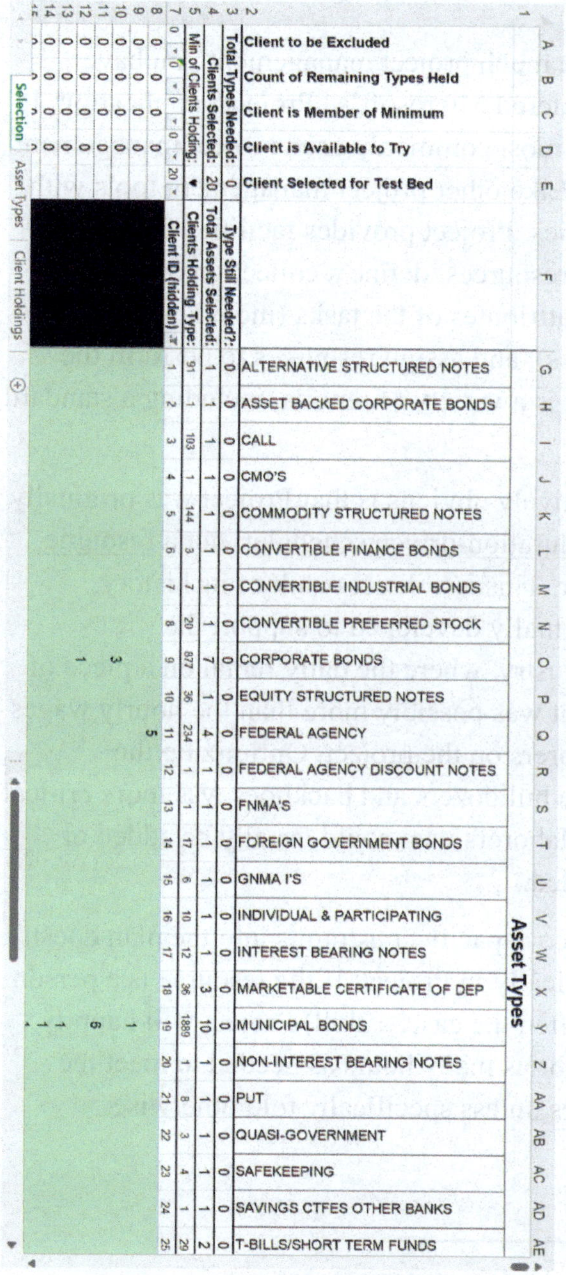

Heavy Calculations: Project Scheduling

If you have done much project management you have probably encountered Microsoft's "Project" application. It is the industry's most commonly used project management software tool[71]. Like other project management tools with similar capabilities, Project provides facilities for the user to enter a set of resources, define a collection of tasks, specify various attributes of the tasks (including duration and dependencies), and assign resources to perform the tasks. The result is a variety of reports, including a standard Gannt chart.

What isn't intuitively obvious is that Project was originally developed as a duration-driven scheduler, not a resource-driven scheduler. This is a direct result of its history. Project was originally developed to support the construction industry, where the daily rental on a piece of heavy equipment was possibly more than the hourly wages of all of the laborers on the project. Optimizing the utilization of the bulldozers and backhoes was more critical than that of the laborers who could usually be added or dropped as needed.

Building resource capacity constraints into the plan doesn't happen automatically in Project. If the resource is a person (or as is more often the case, a skill) Project will happily schedule them for as many hours as needed to meet the project deadlines unless specifically told otherwise.

---

[71] Be nice. I didn't say it was the best.

While this approach might work well for a construction project, it's not so good in other situations. A good chunk of my career has been spent managing technology organizations. The defining characteristic of these organizations is that the resources have specific capabilities (a database administrator and a web developer are not interchangeable) and their availability is limited and relatively fixed (adding a resource is costly and time-consuming). Another characteristic of these organizations is that the demand for services almost always exceeds the ability of the organization to deliver. The critical success factor in this situation is to optimize the utilization of the available resources to maximize delivery and minimize under-utilization. Doing this effectively requires a resource scheduling model that can be easily updated to reflect the current service demands, priorities and capacity constraints.

Recognizing the limitations of Project (not to mention the cost per copy), I determined I needed an effective resource scheduling model.

The output the model would create was obvious – a Gannt chart. Like any Gannt chart, the rows would show the tasks, the task-specific scheduling parameters, and the resource assigned, the columns would show the scheduling periods, and each cell would show the amount of the assigned resource consumed performing that task in that period. Using conditional formatting, each cell that had a value greater than zero would be highlighted, giving the appearance of a bar. Sounds easy!

Well, not so much. The essence of the scheduling problem is determining how all the various parameters affect the value in each of those cells. The secret to solving the scheduling problem is recognizing that scheduling is a problem of finding a minimum: Each parameter represents a constraint that limits the amount of resource that can be used on a task. The amount of resource that can be scheduled is the minimum of all applicable constraints.

As with many problems, the first step to solving the scheduling problem is defining the appropriate units of measure. Unlike many problems, in project management money is typically an output, not an input. The resulting schedule is applied to the unit cost of each resource and summed to determine the total project cost. For the reasons discussed in Chapter 6 the unit of measure should be as granular as the characteristics of the problem reasonably allow.

Each "cell" in the Gannt chart represents the quantity of resource consumed by a particular task during a specific period. When I developed the project scheduling model a key objective was that it could be used by me for overall organizational planning as well as by my project managers to run their individual projects. Units of measure must be defined that works for each of these and meet both of these requirements:

- **Period** – The unit of measure chosen for time periods was hours. This provided flexibility to produce schedules in multiple timeframes. If a weekly schedule was desired, each period had 40 hours. For a monthly schedule, the number of hours in the period depended on the number of days in the month and where the weekends and holidays fell. The model included a calendar that listed the company holidays and used the

NETWORKDAYS function to calculate an accurate number of hours in each month. This same approach was used to determine the number of hours in a quarter. Planning based on hours also allowed the model to feed directly into the budgeting and billing systems.

- **Task** – The project management literature describes many ways to define project plans. Our organization prepared project plans using a three-level hierarchy – phase, activity, task. The project management standards provided a definition for each level. Resources were always assigned at the task level. For simplicity, each project was assigned an internal numeric project number. A "project table", keyed on the internal project number, held the project description, billing codes, and other information used in the management reports. The phase/activity/tasks were identified as decimal values within the project number.
- **Resource** – Our resources were managed as pools, with each pool representing a specific skill set. For planning purposes, the tasks in each project plan would specify the pool (i.e. skill set) required. Once the overall schedule was laid out specific individuals would be assigned to the projects based on their availability. For simplicity, the pools were simply numbered. A "resource table", keyed on the resource pool number, held the description of the skill set the pool represented and the number of resources in the pool each period. A fringe benefit of this approach was that the size of each pool could grow or shrink as resources were hired or moved.

The formula in each cell of the Gannt chart was written so that the calculation hierarchy flowed from left to right and top to bottom. Since the columns represented the scheduling periods, each successive period was calculated after the previous period. Since the rows represented the projects and their tasks, the higher the row (i.e. lower row number) the higher the project's relative priority, and it would have resources allocated to it before a lower priority project. Since the project plans all referenced the internal project number, changes to the relative priority of a project could be reflected by simply assigning it a different number and sorting the task list.

The overall layout of the Gannt tab was as follows:

- The rows in the top section of the tab held the period data. Each column represented a scheduling period. These rows had the relative period number (1 through 48), the number of hours in the period (which would vary depending on how the model was being run, and period beginning date.
- The columns in the left section of the tab (columns A through U) held the project plan data and scheduling parameters. Each row had the project/phase/activity/task number and description. The scheduling parameters included the total estimated hours for the task, the resource type (pool) required, maximum number of resources that could be assigned, maximum percentage of any resource that could be allocated, task dependency (if any) and earliest start date (if any).
- The center section of the tab held the resource allocation calculations. Each cell in this section had the same formula. When a formula was updated to correct an issue or to add a feature it was simply copied to all the other cells in the section.

The "magic" is in the allocation formula. An example of the formula (for the cell in column Z, row 120) is shown below in Figure 24[72].

---

[72] This same formula was shown in Chapter 7 (Figure 13) as an example of how to put comments into a formula.

FIGURE 24. RESOURCE ALLOCATION CALCULATION

```
=
IF(ISTEXT($M120),0+N("Task is on hold"),
IF(Z$2<$M120,0+N("Schedule date is before early start date"),
IF(ISNUMBER($Q120)+N("Resource has been assigned to this task"),
ROUND(MAX(0,
MIN(
IF(ISNUMBER($P120),$P120,99999)+N("Maximum allowed task hours"),
IF(ISNUMBER($O120),$O120*Z$3,99999)+N("Maximum allowed FTE"),
IF(ISNUMBER($N120),$N120*VLOOKUP($Q120,Avail_Hrs,Z$1+3),99999)+N("Maximum allowed percent"),
IF(ISNUMBER($S120),$S120-SUM($V120:Y120),99999)+N("Task hours remaining"),
BG120+N("Resource available hours"),
IF(ISNUMBER($L120),VLOOKUP($L120,DepTbl,Z$1+1),99999)+N("Task dependency constraint")
)),0),0)))
```

How each element of the formula contribute to determining the resource schedule is as follows:

- **IF(ISTEXT($M120),0** – The parameter in Column M is an "Early Start Date". This was used when it was necessary to coordinate our activities with other organizations. If the field contains a date no resources will be scheduled until at least that date. If the field contains "HOLD" (or some other text string) the task won't be scheduled.
- **IF($Z2<$M120,0** – Row 2 has the start date for each period. If this is less than the "Early Start Date" no resources will be scheduled in that period.
- **IF(ISNUMBER($Q120)** – Column Q has the ID of the resource assigned to the task. If there wis nothing in that cell then either a resource has not been assigned or it isn't a task row.
- **ROUND(MAX(0,** – The number of hours scheduled must always be a whole number greater than or equal to zero. Some of the subsequent calculations could yield a number less than zero. This prevents that.

- **MIN(** – As discussed above, scheduling involves finding the minimum of the factors that impact the schedule. These factors are the following. Note that in each of these factors if the constraint doesn't apply the condition returns 99999 hours. Since this is larger than any value the applied factors can return it won't impact the final value.
- **IF(ISNUMBER($P120),$P120,99999)** – If specified, the parameter in Column P has the maximum number of hours that can be allocated to a task each period.
- **IF(ISNUMBER($O120),$O120*Z$3,99999)** – If specified, the parameter in Column O has the maximum number of FTEs that can be allocated to a task each period. If specified, the maximum number of hours that can be allocated is that number times the total number of hours in the period (in row 3)
- **IF(ISNUMBER($N120),$N120*VLOOKUP($Q120, AVAIL_HRS,Z$1+3,0),99999)** – If specified, the parameter in Column N has the maximum percent of the resource's time that can be allocated to a task in each period. The AVAIL_HRS table has the avail hours by period for each resource, net of administrative overhead and MOB requirements.
- **IF(ISNUMBER($S120),$S120-SUM($V120:Y120), 99999)** – If specified, the parameter in Column S is the number of resource hours the task will require in total. The SUM of Columns V to Y gives the total resource hours allocated up to the previous period. The difference is what is left to allocate. Note that as the formula is copied across each additional period is included in the SUM.
- **BG120** – To the right of the Gannt table is a set of calculations with the same dimensions as the Gannt table that has the total number of hours in each period still available for the resource assigned to the task (i.e.: not already allocated to a higher priority task). This is

determined using a SUMIF with the resource number as the criteria.

- **IF(ISNUMBER($L120),VLOOKUP($L120,DepTbl, Z$1+1),99999)** – If specified, the parameter in Column L has a project/phase/activity/task number of a task that the current task (i.e. the one in row 120) is dependent on. In other words, this task cannot start until the other task has completed. The DepTbl table has the maximum number of hours by task by period that any task dependent on that task can schedule in each period. This is determined by comparing the amount of effort remaining to be allocated to the number of hours in the period. If the remaining effort is greater than the hours in the period then the amount that any dependent task can allocate is zero. If it is less than the hours in the period then the amount is the difference.
- **)),0),0)))** – Correctly close all the conditionals

Figure 25 below show a high level view of the upper left hand corner of the Gannt chart.

FIGURE 25 RESOURCE SCHEDULE / GANNT CHART

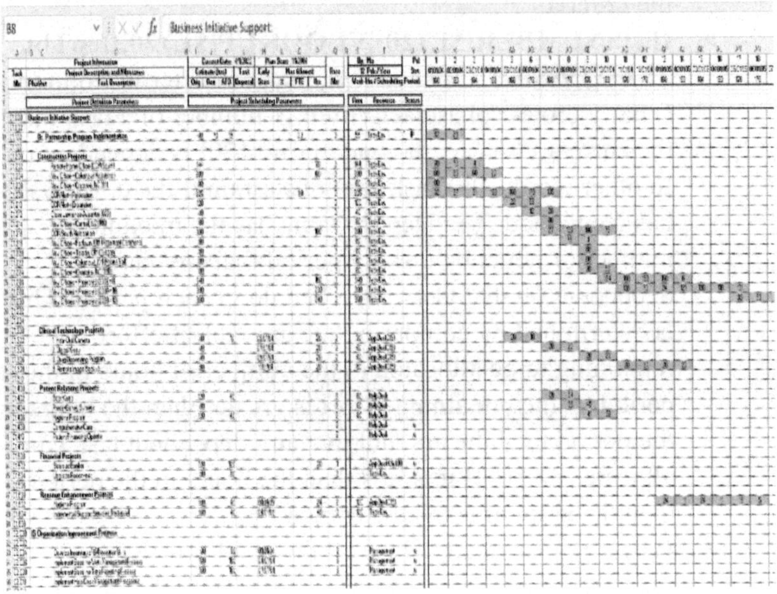

Working Backwards: Organization Chart

There are many problems which can be either easy or difficult depending on which way you approach them. To demonstrate this I sometimes tell people the following:

"I've always been interested in genealogy. I've been able to trace my lineage all the way back to the early 1500s. That's about 20 generations. Looking at my family tree I've noticed something which seems peculiar – In every generation, regardless of how many or how few children the family had, there was always at least one male child. What are the odds of that happening?"

Almost always when I say this they are impressed, both at the presumed rarity of this result and the amount of work I must have done to make this determination. Their thinking usually goes along these lines:

"The odds of a child being male or female is roughly .5 (50%). So over 20 generations that would be (the average number of children per generation) times .5^20. I don't know what that number is, but I know it's really small."

But instead of looking at the problem from top down (i.e., 1500 to today) look at it from the other direction. When viewed this way the problem becomes:

"Did every generation in my family tree have a father?"

Well, Duh! It doesn't matter if I go back 20 generations or 20,000. The answer is yes. This applies to me and every individual in every species that engages in sexual reproduction. The answer is always yes.

The same can hold true for spreadsheets. Sometimes a problem is easier to address if one starts with the final result and works backwards: Given the result, what structure is needed to yield that result? What do I have to do to create that structure? How do I get there with the data I have available?

One of the challenges faced by every significant project is identifying everyone who is going to be affected. In the case of one project I led this would eventually be everyone in the company. Who and when was based on how we scheduled the implementation.

To aid our implementation planning we needed a comprehensive organization chart. Unfortunately, one didn't exist at the level of detail we required. What did exist was an HR system that identified every employee (all 70,000 of them) and the name of their administrative manager. The problem was turning the bottom-up data we had available into a top-down representation.

We decided that a Pivot Table would be the ideal format, since we could use it to expand or collapse any level in any area as needed. The format of the Pivot Table was simple:

- Level 01 – The CEO
- Level 02 – All the executives who reported to the CEO
- Level 03 – All the managers who reported to them.
- Level 04 – Etc.

After some trial-and-error we determined that the longest "hierarchy chain" in the company had 10 levels. We also realized that we couldn't use the person's title as a surrogate for level – the only way to accurately determine where someone belonged was by climbing the chain.

So, how to create a table with the appropriate levels identified for each employee? Once we figured the process out it was actually quite straight-forward:

- Create a table with the name of every employee and the name of their manager. A dummy entry was added to the table for the CEO reporting to himself.
- Add a column to the table with the name of the person to whom the manager reported by looking up the manager's name in the table.
- Add another column in the same way for the manager's manager. Repeat ten times. This assured that the last column in the table for every employee's entry had the name of the CEO.
- Create ten more columns. Working backwards, each column was concatenated to the previous column with the name of the previous manager if different, and a separator character ("|"). If the name of the previous manager was the same as the current manager, then there was no change.
- The last of the ten additional columns had the names of each person in the full reporting hierarchy, from the CEO down, separated by "|", with the employee's name last.
- Create ten more columns with the position of the separator characters as determined by successive FIND commands. A "#VALUE" meant the end of the hierarchy had been reached,
- Create ten more columns, labeled LEVEL01 to LEVEL10. Each column had the name of an employee extracted from the "Master" Column using the values from the FINDs in the previous ten columns. All the columns past the end of the hierarchy were null.
- The table (70,000 rows by 41 columns) was used as the source for a Pivot Table. The row specification of the Pivot Table were the columns labeled Level01, Level02,

Level03, etc. There were no columns or row sums specified.

The beauty of this approach was that we could update the spreadsheet every day from an extract of the HR system and we always had the most current information. With this we could always be assured we were contacting all of the affected employees and their managers as we proceeded with the rollout.

Just for Fun: Sudoku[73]

OK. I'll admit it – I'm a spreadsheet junkie. I create spreadsheets just for the fun of it. Often when I look at a problem I think about how it could be solved using a spreadsheet.

My brother is a Sudoku junkie. Once when I was visiting him he explained that there are two basic strategies in Sudoku:

- **Strategy 1**: Look at each open cell and determine how many different numbers can go into that cell. If there is only one valid number, put that number in the cell and repeat.
- **Strategy 2**: For each number (1 – 9) determine which cells that number can be put into. If there is a 3X3 square where there is only one cell that the number can go into then put the number into that cell and repeat.

Typically one has to cycle between these strategies when solving a puzzle. In more difficult puzzles these strategies might not always yield a unique value. So there may be some trial-and-error involved.

Never one to shy away from a challenge, I of course immediately asked myself: How do you solve a Sudoku puzzle with a spreadsheet?

As with most problems, The first question is "what data do I have to work with?". A Sudoku puzzle consists of a 9X9 square divided into 9 3X3 squares. At the beginning of the

[73] I'm not going to explain Sudoku puzzles here. If you are not familiar with the game look it up and come back later.

game some of the squares are filled in for you. You have to fill in the rest. When you do the Sudoku puzzle in the Sunday paper you write your trial values right in the given square. It is recommended that you do this in pencil as you may discover that a value you try doesn't always work.

Conceptually then a Sudoku puzzle has three types of data:

- The given values
- The values I would supply
- The combined values of the other two squares

I chose to represent the data in three separate 9X9 squares to make them easier to visualize.

FIGURE 26. SUDOKU DATA 1

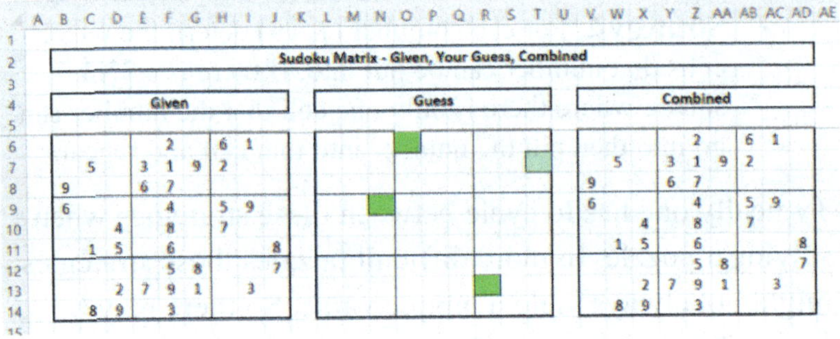

The next challenge was to develop a way to present the results of the two strategies. Given the graphic nature of the problem, the obvious choice was to present them in 9X9 squares. Each square would show the test values identified by the strategy, and the "Guess" square would show where the value would be placed. I used conditional formatting to highlight the appropriate cell in each of the 9X9 squares. I used different colors so I could tell which strategy was suggesting what value.

FIGURE 27. SUDOKU DATA 2

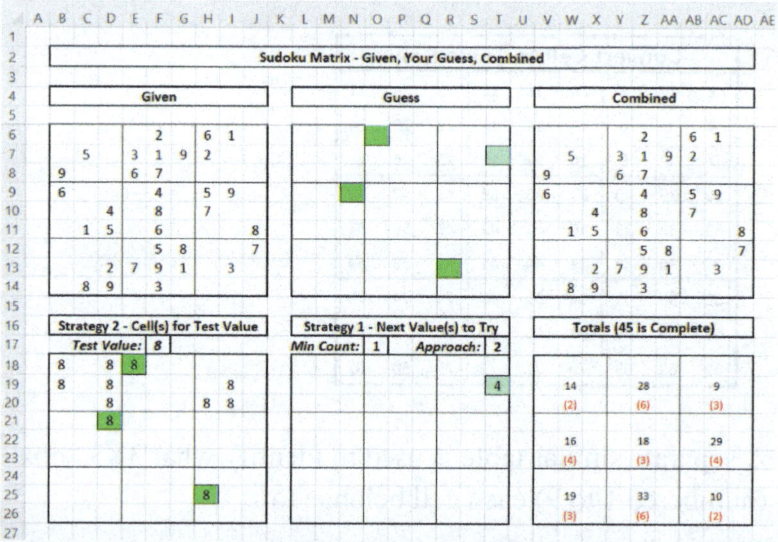

(The third square on the bottom row shows a running total of the values in each Combined 3X3 square and a count of how many cells have been filled.)

Now for the hard part: I know what should be in each of the squares; how to get them there.

The essence of this problem is testing, specifically, testing values to see if they will work in any given cell. Excel is certainly capable of performing a variety of tests. The best way to do this is to create a two-dimensional table with each row representing an individual cell and the columns of the table being the various values involved in the tests. The problem here is that the puzzle is inherently three-dimensional (rows, columns, values). This is handled by creating a 9X9 square that maps each cell in the puzzle to a number from 1 to 81. This row number is used to lookup values in table with the tests.

FIGURE 28. SUDOKU DATA 3

Convert Cell to Table Row								
1	2	3	10	11	12	19	20	21
4	5	6	13	14	15	22	23	24
7	8	9	16	17	18	25	26	27
28	29	30	37	38	39	46	47	48
31	32	33	40	41	42	49	50	51
34	35	36	43	44	45	52	53	54
55	56	57	64	65	66	73	74	75
58	59	60	67	68	69	76	77	78
61	62	63	70	71	72	79	80	81

A separate similar table is used to identify what 3X3 "box" (numbered 1 to 9) each cell belongs to.

FIGURE 29. SUDOKU DATA 4

Convert Cell to a Box Number								
1	1	1	2	2	2	3	3	3
1	1	1	2	2	2	3	3	3
1	1	1	2	2	2	3	3	3
4	4	4	5	5	5	6	6	6
4	4	4	5	5	5	6	6	6
4	4	4	5	5	5	6	6	6
7	7	7	8	8	8	9	9	9
7	7	7	8	8	8	9	9	9
7	7	7	8	8	8	9	9	9

The heart of the spreadsheet is the test table. The test table contains the data from the 9X9 Combined squares in a form that makes them easy to use in formulas (i.e. the formulas can be copied without the need for adjustment). The test table has 81 rows, one for each of the cells in the 9X9 squares. The columns of the test table are as follows (refer to the Figure below):

- **Lkup** – The table row number. This value is used in various VLOOKUP functions.
- **Box** – The number of the box (1 to 9).
- **Row** – The actual row of the cell in the Combined square. This value is used in INDIRECT functions.
- **Col** – The actual column of the cell in the Combined square. This value is used in INDIRECT functions.
- **Used** – The value of this cell in the Combined square (determined using INDIRECT).
- **Combined Row Contents** – 9 cells that have the value of each cell in this row of the Combined square (determined using INDIRECT).
- **Combined Col Contents** – 9 cells that have the value of each cell in this column of the Combined square (determined using INDIRECT).
- **Combined Box Contents** – 9 cells that have the value of each cell in the box (3X3 square) of this cell in the Combined square (determined by a VLOOKUP into a separate box table.
- **Allowed** – 9 cells that will have either the number (if that number isn't used anywhere in the row, column or box of this cell) or dash if that number has been used.
- **Test** – For Strategy 2 the user enters a test number and the Strategy 2 9X9 square shows each cell where that number can be placed. The Test column has a value of one if that number can be placed in that cell.
- **Cnt** – The Strategy 1 9X9 square shows each cell where there is the minimum number of possible values. Ideally, the minimum is one, and shows exactly where that number should go. The Cnt column has the count of the numbers that can still be placed in that cell.
- **Try** – For those cells where the number of possibilities is equal to the minimum the Try column shows the value to try.

FIGURE 30. SUDOKU DATA 5

Approach (1=MIN,2=MAX): 2

Lkup	Box	Row	Col	Used	Combined Row Contents (Col)									Combined Column Contents (Row)									Combined Box Contents (Rel Cell)									Allowed (Current)									Test	Cnt	Try
					V	W	X	Y	Z	AA	AB	AC	AD	6	7	8	9	10	11	12	13	14	1	2	3	4	5	6	7	8	9	1	2	3	4	5	6	7	8	9			
1	1	6	V	0	0	0	0	0	2	0	6	1	0	0	0	9	6	0	0	0	0	0	0	0	0	0	5	0	9	0	0	-	-	3	4	-	-	7	8	-	1	4	0
2	1	6	W	0	0	0	0	0	2	0	6	1	0	0	5	0	0	0	1	0	0	8	0	0	0	0	5	0	9	0	0	-	-	3	4	-	-	7	-	-	0	3	0
3	1	6	X	0	0	0	0	0	2	0	6	1	0	0	0	0	0	4	5	0	2	9	0	0	0	0	5	0	9	0	0	-	-	3	-	-	-	7	8	-	1	3	0
4	1	7	V	0	0	5	0	3	1	9	2	0	0	0	0	9	6	0	0	0	0	0	0	0	0	0	5	0	9	0	0	-	-	-	4	-	-	7	8	-	1	3	0
5	1	7	W	5	0	5	0	3	1	9	2	0	0	0	5	0	0	0	1	0	0	8	0	0	0	0	5	0	9	0	0	-	-	-	-	-	-	-	-	-	0	-	0
6	1	7	X	0	0	5	0	3	1	9	2	0	0	0	0	0	0	4	5	0	2	9	0	0	0	0	5	0	9	0	0	-	-	-	-	-	6	7	8	-	1	3	0
7	1	8	V	9	9	0	0	6	7	0	0	0	0	0	0	9	6	0	0	0	0	0	0	0	0	0	5	0	9	0	0	-	-	-	-	-	-	-	-	-	0	-	0
8	1	8	W	0	9	0	0	6	7	0	0	0	0	0	5	0	0	0	1	0	0	8	0	0	0	0	5	0	9	0	0	-	2	3	4	-	-	-	-	-	0	3	0
9	1	8	X	0	9	0	0	6	7	0	0	0	0	0	0	0	0	4	5	0	2	9	0	0	0	0	5	0	9	0	0	1	-	3	-	-	-	-	8	-	1	3	0
10	2	6	Y	0	0	0	0	0	2	0	6	1	0	0	3	6	0	0	0	0	7	0	0	2	0	3	1	9	6	7	0	-	-	-	4	5	-	-	8	-	1	3	0
11	2	6	Z	2	0	0	0	0	2	0	6	1	0	2	1	7	4	8	6	5	9	3	0	2	0	3	1	9	6	7	0	-	-	-	-	-	-	-	-	-	0	-	0
12	2	6	AA	0	0	0	0	0	2	0	6	1	0	0	9	0	0	0	0	8	1	0	0	2	0	3	1	9	6	7	0	-	-	-	4	5	-	-	-	-	0	2	0
13	2	7	Y	3	0	5	0	3	1	9	2	0	0	0	3	6	0	0	0	0	7	0	0	2	0	3	1	9	6	7	0	-	-	-	-	-	-	-	-	-	0	-	0
14	2	7	Z	1	0	5	0	3	1	9	2	0	0	2	1	7	4	8	6	5	9	3	0	2	0	3	1	9	6	7	0	-	-	-	-	-	-	-	-	-	0	-	0
15	2	7	AA	9	0	5	0	3	1	9	2	0	0	0	9	0	0	0	0	8	1	0	0	2	0	3	1	9	6	7	0	-	-	-	-	-	-	-	-	-	0	-	0
16	2	8	Y	6	9	0	0	6	7	0	0	0	0	0	3	6	0	0	0	0	7	0	0	2	0	3	1	9	6	7	0	-	-	-	-	-	-	-	-	-	0	-	0
17	2	8	Z	7	9	0	0	6	7	0	0	0	0	2	1	7	4	8	6	5	9	3	0	2	0	3	1	9	6	7	0	-	-	-	-	-	-	-	-	-	0	-	0
18	2	8	AA	0	9	0	0	6	7	0	0	0	0	0	9	0	0	0	0	8	1	0	0	2	0	3	1	9	6	7	0	-	-	-	4	5	-	-	-	-	0	2	0
19	3	6	AB	6	0	0	0	0	2	0	6	1	0	6	2	0	5	7	0	0	0	0	6	1	0	2	0	0	0	0	0	-	-	-	-	-	-	-	-	-	0	-	0

The rest of the spreadsheet consists of populating the cells in the Strategy 1 and Strategy 2 9X9 squares by looking up the appropriate values in the test table. Conditional formatting is used extensively to make the results more obvious and visually appealing.

Will this spreadsheet solve every Sudoku puzzle?

No.

If you are looking for a computer Sudoku solution there are many available on the internet for both PCs and mobile devices. Everyone I have looked at uses procedural program code. Could macros be added to this spreadsheet to make it more robust?

Certainly.

But that is not why this spreadsheet is being presented. The purpose of this and the other examples in this chapter is to show that there are a wide variety of non-conventional problems that can be addressed using spreadsheets. The first, basic question that must always be asked is:

What is the most appropriate tool to solve the problem?

Appropriate can incorporate many considerations, including efficiency, effectiveness, ease of implementation, and ease of use. If the answer to that question is a spreadsheet, make sure it is built and managed correctly.

11. FINAL THOUGHTS

In many ways, electronic spreadsheets launched the data driven management revolution. Spreadsheet software grew and evolved somewhat in parallel with the large mainframe-based enterprise ERP systems. The ERP systems produced massive amounts of data, and the spreadsheet users eagerly consumed it. Their data appetite demanded more and more data. The enterprise systems happily obliged.

And the spiral has been expanding ever since!

Today we are in the "Data Driven Management" era. Vast quantities of data are generated by complex automated manufacturing, distribution, sales and financial systems. This data is piped directly into a variety of sophisticated management and control systems which generate real-time process adjustments. In many cases these adjustments are fed directly back into the operational systems without any human intervention.

This feedback loop and the systems which implement it drive the tactical management needs of the business. In addition, businesses need systems to support their strategic management needs. These are the systems which input the operational data, transform it into valuable information, and present it in a useful form for human consumption.

Sounds familiar? It should be. This is exactly what a well-designed and well-constructed spreadsheet does. This is not to say that spreadsheets are the only, or even the best, way to do this in every situation. There are many considerations and there are a wide range special-purpose and general-purpose analytics tools available. These tools offer a wide variety of sophisticated capabilities significantly more advanced than those available from a spreadsheet.

One such tool that I have used is Tableau. Tableau is a general purpose "data visualization" tool. It provides a wide range of sophisticated presentation capabilities. These include:

- Ability to input and process data from a wide range of sources and formats ranging from spreadsheets to enterprise relational databases to unstructured "Big Data".
- Continuous automatic updating as the source systems change.
- Intuitive analytics and smart dashboards.
- Integrated interfaces to web applications and mobile devices.
- Speed!

As I discussed in Chapter 2, it is important to select the right tool for the problem. A spreadsheet is not always the right tool. Depending on the nature and amount of data (and other considerations) the problem may be better addressed with a more robust tool such as Tableau.

But the converse is also true: Just because Tableau has superior data presentation capabilities doesn't mean it is the best for every application. Tableau's greatest strength is user-driven data visualization. Where there is a need for sophisticated user-driven data transformation (within the capacity limits) a spreadsheet may still be the best choice.

There are other considerations as well. One is cost. In a practical sense spreadsheets are "free". The widely used Microsoft suite of office productivity tools includes Excel. There are also equivalent and largely compatible products available from Apple, Google and others. Tableau on the other hand is licensed per user per month. And it isn't cheap.

Another consideration is commonality. There are vastly more people who are familiar with spreadsheets than with these other tools. This makes collaboration and collective problem-solving easier and more effective. My own opinion is that this situation is unlikely to change any time soon. New and "better" tools come on the market every year. None of them has yet been the overwhelming game changer that the electronic spreadsheet was when it was introduced.

The Rumors of My Death ...

As I was putting this book together several people made comments to the effect that "the days of the spreadsheet are over". In some respects they are right. The electronic spreadsheet made its debut during the early days of the mainframe era. At that time data was limited and hard to access. And access to that data was limited to the privileged few (i.e., the geeks). The spreadsheet changed that. Suddenly anyone could access and manipulate[74] significant quantities of data. It truly did create an entirely new paradigm.

Now that situation has certainly changed – data is everywhere. In fact, we are drowning in it. And our ability to crunch that data has grown as well. The processor in my cell phone has more capability than the mainframes that were running the world when the spreadsheet was introduced. And the tools and facilities for accessing and using that data today weren't even dreamed of back them. (Yes, Virginia, there was a time before the internet.)

But in other respects I feel they are wrong. While there are many products that can outperform any individual spreadsheet capability, I don't know of any that can outperform all of them.

Will this change at some point? I'm sure it will. That is the nature of progress. Someday someone will have a brilliant idea that will radically change the paradigm again.

---

[74] I know. I promised not to use that word.

What will that look like? I haven't a clue. When will it come? I don't know. I am pretty sure though that it's not going to be any time soon.

Spreadsheets have become well entrenched in our collective organizational culture. And for good reason – they provide immensely valuable services, and collectively does them better than anything else currently available.

And even when the paradigm does shift the underlying requirements will still be the same. Spreadsheets after all are just another tool. To use it effectively you must understand the essential characteristics of the problem and match that against the capabilities of the tools at your disposal.

Then, and only then, can you choose right tool, be it a spreadsheet or a hammer.

APPENDIX A: COMMON EXCEL KEYBOARD SHORTCUTS

(Partial List)

Letters:

Keyboard Combination						
Key	Alt	Ctrl	Shift	Keystrokes	Result	Standard Ribbon Commands if Available
A		✔		Ctrl+A	Select Entire Region	
		✔	✔	Ctrl+Shift+A	Insert Argument Names into Formula	
	✔			Alt+A	Select Data Ribbon Tab In Menu Mode	Data
B		✔		Ctrl+B	Toggle Bold	Home > Format > Cells > Font > Font Style > Bold
C		✔		Ctrl+C	Copy Selected Cells	Home > Copy
D		✔		Ctrl+D	Fill Down	
E		✔		Ctrl+E	Invoke Flash Fill	
E	✔			Alt+E, S	After Copy/Cut: Open Paste Special Diaglog Box	Home > Paste > Paste Special...
F		✔		Ctrl+F	Find	Home > Find & Select
		✔	✔	Ctrl+Shift+F	Open Format Cells Dialog Box	Home > Format Cells
	✔			Alt+F	Select File Ribbon Tab In Menu Mode	File
G		✔		Ctrl+G	Goto	Home > Find & Select > Goto
		✔	✔	Ctrl+Shift+G	Open Workbook Statistics Dialog Box	
H		✔		Ctrl+H	Find & Replace	
	✔			Alt+H	Select Home Ribbon Tab In Menu Mode	Home
I		✔		Ctrl+I	Toggle Italic	Home > Format > Format Cells > Font > Italic
J	✔			Alt+J	Select Draw Ribbon Tab In Menu Mode	Draw
K		✔		Ctrl+K	Open Insert Hyperlink Dialog Box	Insert > Links
L		✔		Ctrl+L	Open Create Table Dialog Box	
M	✔			Alt+M	Select Formulas Ribbon Tab In Menu Mode	Formulas
N		✔		Ctrl+N	New Workbook	File > New
	✔			Alt+N	Select Insert Ribbon Tab In Menu Mode	Insert
O		✔		Ctrl+O	Open A Workbook	File > Open
		✔	✔	Ctrl+Shift+O	Select All Cells That Contain Comments	
P		✔		Ctrl+P	Print	File > Print
		✔	✔	Ctrl+Shift+P	Open Format Point Size Drop Down List	Home > Format > Format Cells > Font > Size
	✔			Alt+P	Select Page Layout Ribbon Tab In Menu Mode	Page Layout
Q		✔		Ctrl+Q	Display Quick Analysis Options	
	✔			Alt+Q	Open Search Menu	
R		✔		Ctrl+R	Fill Right	
	✔			Alt+R	Select Review Ribbon Tab In Menu Mode	Review
S		✔		Ctrl+S	Save Workbook	File > Save
T		✔		Ctrl+T	Open Create Table Dialog Box	
U		✔		Ctrl+U	Toggle Underline	Home > Format > Format Cells > Font > Underline > Single
		✔	✔	Ctrl+Shift+U	Expand/Collapse Formula Bar	
V		✔		Ctrl+V	Paste	Home > Paste
	✔	✔		Alt+Ctrl+V	After Copy/Cut: Open Paste Special Diaglog Box	Home > Paste > Paste Special...
W		✔		Ctrl W	Close Workbook	File > Close
	✔			Alt+W	Select View Ribbon Tab In Menu Mode	
X		✔		Ctrl+X	Cut	Home > Cut
Y		✔		Ctrl+Y	Repeat Last Action	
Z		✔		Ctrl+Z	Undo Last Action	

Numbers:

Keyboard Combination						
Key	Alt	Ctrl	Shift	Keystrokes	Result	Standard Ribbon Commands if Available
1		✔		Ctrl+1	Open Format Cells Dialog Box	Home > Format > Format Cells
		✔	✔	Ctrl+Shift+1 [!]	Apply Default Number Format	
2		✔		Ctrl+2	Toggle Bold	Home > Format > Format Cells > Font > Bold
		✔	✔	Ctrl+Shift+2 [@]	Apply Default Time Format	Home > Format > Format Cells > Number > Time
3		✔		Ctrl+3	Toggle Italic	Home > Format > Format Cells > Font > Italic
		✔	✔	Ctrl+Shift+3 [#]	Apply Default Date Format	Home > Format > Format Cells > Number > Date
4		✔		Ctrl+4	Toggle Underline	Home > Format > Format Cells > Font > Underline > Single
		✔	✔	Ctrl+Shift+4 [$]	Apply Default Currency Format	Home > Format > Format Cells > Number > Currency
5		✔		Ctrl+5	Toggle Strikethrough	Home > Format > Format Cells > Font > Strikethrough
		✔	✔	Ctrl+Shift+5 [%]	Apply Default Percent Format	Home > Format > Format Cells > Number > Percent
	✔	✔		Alt+Ctrl+5	Select Floating Objects	
6		✔		Ctrl+6	Toggle Show/Hide Objects	
		✔	✔	Ctrl+Shift+6 [^]	Apply Exponential Format	Home > Format > Format Cells > Number > Scientific
7		✔	✔	Ctrl+Shift+7 [&]	Place Outline Border Around Selected Cells	Format > Cells > Border
8		✔		Ctrl+8	Toggle Outline Symbols	
		✔	✔	Ctrl+Shift+8 [*]	Select Region Around The Active Cell	
9		✔		Ctrl+9	Hide Rows	Home > Format > Hide & Unhide > Hide Rows
		✔	✔	Ctrl+Shift+9 [(]	Unhide Rows	Home > Format > Hide & Unhide > Unhide Rows
0		✔		Ctrl+0	Hide Columns	Home > Format > Hide & Unhide > Hide Columns
		✔	✔	Ctrl+Shift+0 [)]	Unhide Columns	Home > Format > Hide & Unhide > Unhide Columns

FunctionKeys:

Key	Alt	Ctrl	Shift	Keystrokes	Result	Standard Ribbon Commands if Available
F1				F1	Help	Help > Help
		✔		Ctrl+F1	Expand/Collapse Ribbon	
	✔			Alt+F1	Insert Chart	Insert > Charts (Select Chart Type)
	✔		✔	Alt+Shift+F1	New Worksheet	
F2				F2	Edit Cell Contents	
			✔	Shift+F2	Insert/Edit Cell Comment	Insert > Comments
	✔			Alt+F2	Save As	File > Save As
	✔		✔	Alt+Shift+F2	Save	File > Save
		✔	✔	Ctrl+Shift+F2	Insert/Edit Threaded Comment	
F3				F3	Open Paste Name Dialog Box	
			✔	Shift+F3	Open Insert Function Dialog Box	
		✔		Ctrl+F3	Open Name Manager Dialog Box	Formulas > Name Manager
		✔	✔	Ctrl+Shift+F3	Open Create Names Dialog Box	
F4				F4	Cycle Through Absolute/Relative Cell Reference	
			✔	Shift+F4	Find Next	Home > Find & Select > Find > Find Next
		✔		Ctrl+F4	Close	File > Close
	✔			Alt+F4	Exit	File > Exit
F5				F5	Goto	Home > Find & Select > Goto
			✔	Shift+F5	Open Find & Replace Dialog Box	Home > Find & Select > Find
		✔		Ctrl+F5	Restore Window Size	
F6				F6	Next Pane	
			✔	Shift+F6	Previous Pane	
		✔		Ctrl+F6	Next Workbook Window	View > Switch Windows
		✔	✔	Ctrl+Shift+F6	Previous Workbook Window	View > Switch Windows
F7				F7	Open Spell Check Dialog Box	Review > Spelling
		✔		Ctrl+F7	Move Window	
F8				F8	Toggle Extend Mode	
			✔	Shift+F8	Add Range To Current Selection	
		✔		Ctrl+F8	Resize Window	
	✔			Alt+F8	Open Macro Dialog Box	View > Macros
F9				F9	Recalculate All Worksheets In All Open Workbooks	Formulas > Calculate Now
			✔	Shift+F9	Recalculate Active Worksheet	Formulas > Calculate Sheet
		✔		Ctrl+F9	Minimize Workbook	
	✔	✔		Alt+Ctrl+F9	Recalculate All Worksheets In All Open Workbooks	Formulas > Calculate Now
	✔	✔	✔	Alt+Ctrl+Shift+F9	Recheck Dependencies & Recalculate All Open Workbooks	
F10				F10	Activate Ribbon Menu Mode	
			✔	Shift+F10	Display Context Sensitive Shortcut Menu	
		✔		Ctrl+F10	Maximize Or Restore Window	
	✔			Alt Or F10	Open Object Selection Pane	Home > Find & Select > Selection Pane
	✔		✔	Alt+Shift+F10	Display Error Message	
F11				F11	Create Chart	Insert > Chart
			✔	Shift+F11	New Worksheet	Insert > Worksheet
		✔		Ctrl+F11	Insert Macro Sheet	
	✔			Alt+F11	Open Visual Basic Editor	View > Macros > View Macros > Edit
F12				F12	Save As	File > Save As
			✔	Shift+F12	Save	File > Save
		✔		Ctrl+F12	Open A Workbook	File > Open
		✔	✔	Ctrl+Shift+F12	Print	File > Print

Punctuation:

Key	Alt	Ctrl	Shift	Keystrokes	Result	Standard Ribbon Commands if Available
-		✔		Ctrl+-	Open Delete Dialog Box	
		✔	✔	Ctrl+Shift+- [_]	Remove Outline Border	Home > Format > Format Cells > Border > None
'		✔	✔	Ctrl+Shift+' ["]	Copy Value From Cell Above	
/		✔		Ctrl+/	Select The Array Containing The Active Cell	
;		✔		Ctrl+;	Insert Current Date	
		✔	✔	Ctrl+Shift+; [:]	Insert Current Time	
	✔			Alt+;	Select The Visible Cells In The Current Selection	
[		✔		Ctrl+[	Select Cells Directly Referenced By The Active Cell	Formulas > Trace Precedents
		✔	✔	Ctrl+Shift+[[{]	Select Cells Directly Or Indirectly Referenced By The Active Cell	
\		✔		Ctrl+\	Select Cells In Row That Don'T Match The Active Cell	
		✔	✔	Ctrl+Shift+\ [|]	Select Cells In Column That Don'T Match The Active Cell	
]		✔		Ctrl+]	Select Cells That Directly Reference The Active Cell	Formulas > Trace Dependents
		✔	✔	Ctrl+Shift+] [}]	Select Cells That Directly Or Indirectly Reference The Active Cell	
`		✔		Ctrl+`	Toggle Value/Formula Display	
		✔	✔	Ctrl+Shift+` [~]	Apply General Number Format	Home > Format > Format Cells > Number > General

Keyboard Combination						
Key	Alt	Ctrl	Shift	Keystrokes	Result	Standard Ribbon Commands if Available
Miscellaneous:						
<Backspace>			✓	Shift+Backspace	Select Only The Active Cell Within Selected Range	
<Delete>				Delete	Remove Cell Contents	
<Dn Arrow>	✓			Alt+Dn Arrow	Open Autocomplete Dialog Box	
<End>		✓		Ctrl+End	Goto Last Cell In Worksheet	
<Enter>				Enter	Perform The Command	
		✓		Ctrl+Enter	Fill Selected Range With Active Cell Contents	
	✓			Alt+Enter	Start A New Line In The Same Cell	
			✓	Shift+Enter	Select Cell Above Current Cell	
<Home>		✓		Ctrl+Home	Goto First Cell In Worksheet	
<Pg Dn>	✓			Alt+Pg Dn	Move One Screen Right	
		✓	✓	Ctrl+Shift+Pg Dn	Select Current & Next Worksheet	
		✓		Ctrl+Pg Dn	Goto Next Worksheet	
<Pg Up>	✓			Alt+Pg Up	Move One Screen Left	
		✓	✓	Ctrl+Shift+Pg Up	Select Current & Previous Worksheet	
		✓		Ctrl+Pg Up	Goto Previous Worksheet	
<Space>			✓	Shift+Space	Select Entire Row	
		✓		Ctrl+Space	Select Entire Column	
		✓	✓	Ctrl+Shift+Space	Selects All Objects	
	✓			Alt+Space	Open The Window Control Menu	
<Tab>				Tab	Select Next Cell Ir Option	
			✓	Shift+Tab	Select Previous Cell Or Option	
		✓		Ctrl+Tab	Activate Next Worksheet	
		✓	✓	Ctrl+Shift+Tab	Activate Previous Worksheet	
<Alt>	✓			Alt	Activate Ribbon Menu Mode	

APPENDIX B: BASIC DISPLAY FORMAT CODES

Format Specification Structure –

Excel display format codes are specifically intended to make the meaning of numbers clearer and to make the numbers more understandable. Excel provides predefined codes that support several common conventions such as inserting commas to represent thousands, defining the number of decimal places, displaying negative numbers in red or enclosing them in parentheses, etc. There are also predefined codes to help interpret numbers that have a specific meaning. For example, Excel stores dates and times as numbers. If a cell is identified as containing a date or a time Excel applies format codes to display the cell's value in a format more meaningful to the user.

Users can also define their own format codes. The structure of a format code has four sections separated by semi-colons (";"). Each section specifies how a cell that would normally contain a number is to be displayed under different conditions:

<if positive>;<if negative>;<if zero>;<if text>

All four sections do not need to be specified. If only one section is supplied the code applies to all values in the field. If two sections are supplied the first applies to positive numbers and the second to negative numbers.

Format Specification Application –

Application of the format codes is actually a two step process:

- **Step 1** – Excel determines what the number <u>means</u>, based on the codes that have been specified. Be careful! As discussed in Chapter 4, Excel might interpret a number different from what was intended.
- **Step 2** – Once the meaning of the number has been determined, Excel decides how best to present it.

For example, if a cell contains the value 1234.56, and you tell Excel it is a number (or don't tell it anything), it will probably display as 1,234.56. If on the other hand you tell Excel it is a date, it could display as Monday May 18, 1903 (or equivalent). If you tell Excel it is a time, it would be 1:26:24PM.

Format Codes –

Each section of the format specification may contain several different types of codes which control different aspects of the final display. These elements are as follows:

- **Color** – The desired color is specified in brackets. There are a total of 56 colors that can be specified. These can be specified as [color<n>], where <n> is a number from 1 to 56. The first 8 are the most common colors and have names assigned:
  [color1] or [Black]
  [color2] or [White]
  [color3] or [Red]
  [color4] or [Green]
  [color5] or [Blue]
  [color6] or [Yellow]
  [color7] or [Magenta]
  [color8] or [Cyan]

- **“Native” Characters** – These characters can be placed in a format specification and will appear as is:
  $ – Dollar
  +- – Plus / Minus
  () – Parentheses
  {} – Braces
  <> – Less / Greater
  = – Equal
  : – Colon
  ^ – Caret
  ‘ – Apostrophe
  / – Slash
  ! – Exclamation
  & – Ampersand
  ~ – Tilde
    – Space
- **“Non-native” Characters** – Because these characters have special meaning they must be preceded with an escape character (“\”) if they are going to be displayed:
  # – Entered as “\#”
  * – Entered as “*”
  % – Entered as “\%”
- **Numeric Placeholders** –
  0 – Display non-significant digits
  # – Display significant digits
  ? – Align decimal places
  . – Decimal point
  , – Thousands separator
  * – Repeat the character following the asterisk
  _ – Add space (often used to align positives & negatives)
  @ – Text
- **Text Code** – A text string can be displayed following a format code by enclosing it in quotes (“”)
- **Date Format Codes** –
  m – Single digit month number (January to September)
  mm – Two digit month number

mmm – Three character month name abbreviation
mmmm – Full month name
mmmmm – First letter of month name
d – Single digit day number
dd Two digit day number
ddd – Three digit day of the week abbreviation
dddd – Full day of the week name
yy – Two digit year
yyyy – Four digit year

- **Time Format Codes** –

  h,m,s – Single digit hour, minute, second numbers
  hh,mm,ss – Two digit hour, minute, second numbers
  AM / PM – Morning or afternoon
  a / p – Single character morning or afternoon indicator

Note: If an illegal format code is specified Excel will interpret it as best it can; Excel will not give an error message.

APPENDIX C: COMMON ASCII CHARACTER CODES

Everything inside a computer is stored as a binary number. In other words, it is comprised of a fixed number of binary digits, each of which can have a value of zero or one. The number of digits is determined by the type of hardware.

In the early days of computing that's where the story ended. The values in the various registers and memory locations could be displayed as lights on the control panel. This tradition continued even into the "third generation" computers in the 1960s and 1970s.

FIGURE 31. UNIVAC 1108 CONTROL PANEL, CIRCA 1968

It became obvious early on that binary wasn't a useful way for computers to communicate with people. To address this deficiency the computer engineers came up with a system for associating characters (letters, digits, punctuation, etc.) with specific binary values. These "Information Interchange Codes" have proven to be very successful and are still used. The two most common codes used today are:

- **EBCDIC** – Extended Binary Coded Decimal Interchange Code. Used in IBM mainframe computers.
- **ASCII** – American Standard Code for Information Interchange. Used everywhere else, including the personal computers where electronic spreadsheets live.

So why should we care?

Well, the short answer is usually you don't. The most impressive achievement of modern computer systems is that the users, and even the programmers (sorry, developers), usually don't need to understand how the computer actually does its magic. Systems today consist of multiple "layers of abstraction" that (at the top level) provide a "user friendly" interface (if you'll forgive the cliché) where the input is entered and then passed down through successively complex layers until it reaches (at the bottom layer) the bits and bytes of the actual machine instructions.

That said, it is valuable to understand, at least to some degree, what is going on under the covers. This can be very helpful in figuring out the cause of some weird error condition.

How does this relate to Excel? Excel provides functions that allow the user to see and manipulate the ASCII codes that lie underneath the characters that you see on the screen. The CODE(<text>) function returns the ASCII decimal code of the leftmost character in <text>. The CHAR(<decimal value>) function does the opposite. It returns the character whose ASCII code is <decimal value>. These functions can be very useful when it is necessary to manipulate or transform individual characters or create special control sequences for printers or other devices.

Basic ASCII Character Values

Numeric Value					
Binary (Base 2)	Octal (Base 8)	Decimal (Base 10)	Hex (Base 16)	Symbol	Description
Control Characters:					
0000000	000	0	00	NUL	Null char
0000001	001	1	01	SOH	Start of Heading
0000010	002	2	02	STX	Start of Text
0000011	003	3	03	ETX	End of Text
0000100	004	4	04	EOT	End of Transmission
0000101	005	5	05	ENQ	Enquiry
0000110	006	6	06	ACK	Acknowledgment
0000111	007	7	07	BEL	Bell
0001000	010	8	08	BS	Back Space
0001001	011	9	09	HT	Horizontal Tab
0001010	012	10	0A	LF	Line Feed
0001011	013	11	0B	VT	Vertical Tab
0001100	014	12	0C	FF	Form Feed
0001101	015	13	0D	CR	Carriage Return
0001110	016	14	0E	SO	Shift Out / X-On
0001111	017	15	0F	SI	Shift In / X-Off
0010000	020	16	10	DLE	Data Line Escape
0010001	021	17	11	DC1	Device Control 1 (oft. XON)
0010010	022	18	12	DC2	Device Control 2
0010011	023	19	13	DC3	Device Control 3 (oft. XOFF)
0010100	024	20	14	DC4	Device Control 4
0010101	025	21	15	NAK	Negative Acknowledgement
0010110	026	22	16	SYN	Synchronous Idle
0010111	027	23	17	ETB	End of Transmit Block
0011000	030	24	18	CAN	Cancel
0011001	031	25	19	EM	End of Medium
0011010	032	26	1A	SUB	Substitute
0011011	033	27	1B	ESC	Escape
0011100	034	28	1C	FS	File Separator
0011101	035	29	1D	GS	Group Separator
0011110	036	30	1E	RS	Record Separator
0011111	037	31	1F	US	Unit Separator

Basic ASCII Character Values

Numeric Value					
Binary (Base 2)	Octal (Base 8)	Decimal (Base 10)	Hex (Base 16)	Symbol	Description
Regular (Printable) Characters:					
0100000	040	32	20		Space
0100001	041	33	21	!	Exclamation mark
0100010	042	34	22	"	Double quotes (or speech marks)
0100011	043	35	23	#	Number
0100100	044	36	24	$	Dollar
0100101	045	37	25	%	Per cent sign
0100110	046	38	26	&	Ampersand
0100111	047	39	27	'	Single quote
0101000	050	40	28	(	Open parenthesis (or open bracket)
0101001	051	41	29	)	Close parenthesis (or close bracket)
0101010	052	42	2A	*	Asterisk
0101011	053	43	2B	+	Plus
0101100	054	44	2C	,	Comma
0101101	055	45	2D	-	Hyphen
0101110	056	46	2E	.	Period, dot or full stop
0101111	057	47	2F	/	Slash or divide
0110000	060	48	30	0	Zero
0110001	061	49	31	1	One
0110010	062	50	32	2	Two
0110011	063	51	33	3	Three
0110100	064	52	34	4	Four
0110101	065	53	35	5	Five
0110110	066	54	36	6	Six
0110111	067	55	37	7	Seven
0111000	070	56	38	8	Eight
0111001	071	57	39	9	Nine
0111010	072	58	3A	:	Colon
0111011	073	59	3B	;	Semicolon
0111100	074	60	3C	<	Less than (or open angled bracket)
0111101	075	61	3D	=	Equals
0111110	076	62	3E	>	Greater than (or close angled bracket)
0111111	077	63	3F	?	Question mark
1000000	100	64	40	@	At symbol
1000001	101	65	41	A	Uppercase A
1000010	102	66	42	B	Uppercase B
1000011	103	67	43	C	Uppercase C
1000100	104	68	44	D	Uppercase D
1000101	105	69	45	E	Uppercase E
1000110	106	70	46	F	Uppercase F
1000111	107	71	47	G	Uppercase G

Basic ASCII Character Values

Numeric Value					
Binary (Base 2)	Octal (Base 8)	Decimal (Base 10)	Hex (Base 16)	Symbol	Description

Regular (Printable) Characters:

1001000	110	72	48	H	Uppercase H
1001001	111	73	49	I	Uppercase I
1001010	112	74	4A	J	Uppercase J
1001011	113	75	4B	K	Uppercase K
1001100	114	76	4C	L	Uppercase L
1001101	115	77	4D	M	Uppercase M
1001110	116	78	4E	N	Uppercase N
1001111	117	79	4F	O	Uppercase O
1010000	120	80	50	P	Uppercase P
1010001	121	81	51	Q	Uppercase Q
1010010	122	82	52	R	Uppercase R
1010011	123	83	53	S	Uppercase S
1010100	124	84	54	T	Uppercase T
1010101	125	85	55	U	Uppercase U
1010110	126	86	56	V	Uppercase V
1010111	127	87	57	W	Uppercase W
1011000	130	88	58	X	Uppercase X
1011001	131	89	59	Y	Uppercase Y
1011010	132	90	5A	Z	Uppercase Z
1011011	133	91	5B	[	Opening bracket
1011100	134	92	5C	\	Backslash
1011101	135	93	5D	]	Closing bracket
1011110	136	94	5E	^	Caret - circumflex
1011111	137	95	5F	_	Underscore
1100000	140	96	60	`	Grave accent
1100001	141	97	61	a	Lowercase a
1100010	142	98	62	b	Lowercase b
1100011	143	99	63	c	Lowercase c
1100100	144	100	64	d	Lowercase d
1100101	145	101	65	e	Lowercase e
1100110	146	102	66	f	Lowercase f
1100111	147	103	67	g	Lowercase g
1101000	150	104	68	h	Lowercase h
1101001	151	105	69	i	Lowercase i
1101010	152	106	6A	j	Lowercase j
1101011	153	107	6B	k	Lowercase k
1101100	154	108	6C	l	Lowercase l
1101101	155	109	6D	m	Lowercase m
1101110	156	110	6E	n	Lowercase n
1101111	157	111	6F	o	Lowercase o
1110000	160	112	70	p	Lowercase p

Basic ASCII Character Values

Numeric Value					
Binary (Base 2)	Octal (Base 8)	Decimal (Base 10)	Hex (Base 16)	Symbol	Description

Regular (Printable) Characters:

1110001	161	113	71	q	Lowercase q
1110010	162	114	72	r	Lowercase r
1110011	163	115	73	s	Lowercase s
1110100	164	116	74	t	Lowercase t
1110101	165	117	75	u	Lowercase u
1110110	166	118	76	v	Lowercase v
1110111	167	119	77	w	Lowercase w
1111000	170	120	78	x	Lowercase x
1111001	171	121	79	y	Lowercase y
1111010	172	122	7A	z	Lowercase z
1111011	173	123	7B	{	Opening brace
1111100	174	124	7C	|	Vertical bar
1111101	175	125	7D	}	Closing brace
1111110	176	126	7E	~	Tilde
1111111	177	127	7F		Delete

ACKNOWLEDGMENTS

We humans are social creatures. It is not surprising then that most of our worthwhile ventures are the collaborative efforts of many individuals. This book is no exception. In recognition of this, I want to take this opportunity to thank those who helped to make this work possible.

First, I want to thank David Van Horn, an experienced author and publisher. He has provided priceless guidance and encouragement to me and many other fledgling authors. His assistance has been essential in navigating today's complex and ever-changing world of publishing.

To Michael Canales, who's visual magic was able to take my thoughts and turn them into this amazing cover design.

Too numerous to list, the friends and associates who read the drafts and generously pointed out where I had gone too far off the rails.

To the many internet Excel experts who seem to know the answer to every question. (And to Google, who makes them accessible!)

Last, my profound appreciation to my wife and friend Rose, whose patience and support has made it possible for me to tilt at this windmill.

Thank you all.

ABOUT THE AUTHOR

Richard (RB) Weber was born and raised in Washington DC, the son of a career Washington bureaucrat and a stereotypical 1950s stay-at-home housewife. Always an avid reader, he spent much of his youth devouring a wide variety of books and the local newspapers—the Evening Star (now defunct) and the Washington Post (which would probably benefit humanity if it were defunct).

Weber received his primary education courtesy of the Washington DC public school system. At the ripe old age of 18 he graduated from Woodrow Wilson High School (which in a sense is also defunct. The name of the school was recently changed because apparently President Woodrow Wilson was a racist by today's standards for political correctness.).

Being moderately bright but also clueless as to what he wanted to do with his life, he attended Carnegie Mellon University where he received his undergraduate degree in Physics.

Unfortunately, people with physics degrees have limited employment prospects, especially if you're not on the Nobel Prize track. Fortunately, Carnegie also had computers (still does). Between physics labs he learned how to make computers talk, which proved to be far more stimulating than talking to his fellow physics students. This was fortunate, as it enabled him to move to Cleveland,

Ohio after graduation and get a job as a computer programmer.

What followed was a long career in what is now called information technology. Over the years he has held a variety of technical, management and executive positions at several companies in the Cleveland area.

He is now “retired” and living in Houston, Texas. He spends much of his time traveling, writing, consulting and visiting his family (who also live in Houston).

Writing has always been a passion for Mr. Weber. Much of his activities throughout his career has involved writing—from technical specifications to business case analyses to project proposals to strategic plans.

Over the years he has produced a variety of works, both fiction and non-fiction, independent of his professional activities.

If you would like to discuss the material presented in this book or would like an electronic copy, Mr. Weber can be contacted by email at:

WellWornWorks@gmail.com

SOLUTION TO THE PROBLEM ON PAGE 182

This is actually a trick question. Our current calendar system was adopted by decree of Pope Gregory XIII in 1582 (which is why we refer to it as the Gregorian calendar). Prior to that, we (i.e. Europeans) used the Julian calendar (devised by Julius Caesar in 46 BCE), which had an entirely different structure of months. In the Julian calendar, every 4 years was a leap year regardless. After a lot of jiggling to correct previous goofs, the every-4-year rule settled into the groove on 4 CE. So, 900 CE was a leap year (the 225th since 4 CE).

FIGURE 32. THE ORIGINAL COMPUTER BUG

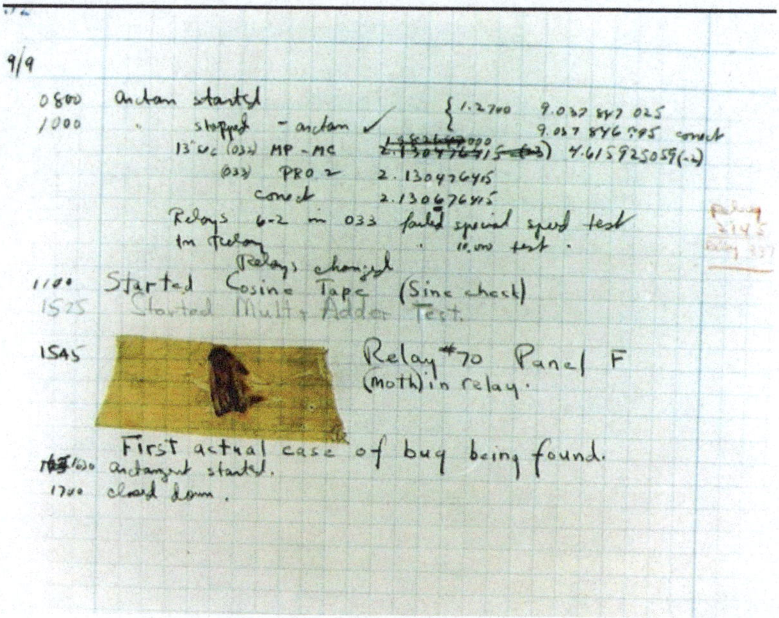

Made in United States
Orlando, FL
04 November 2026

90202168R00174